Green Facilities Handbook

Simple & Profitable Strategies for Managers

Green Facilities Handbook

Simple & Profitable Strategies for Managers

Eric A. Woodroof, Ph.D., CEM, CRM

THE FAIRMONT PRESS, INC.

CRC Press
Taylor & Francis Group

Library of Congress Cataloging-in-Publication Data

Woodroof, Eric A.
 Green facilities handbook : simple & profitable strategies for managers / Eric A. Woodroof.
    p. cm.
  Includes index.
  ISBN 0-88173-607-4 (alk. paper) -- ISBN 0-88173-608-2 (electronic) -- ISBN 1-4200-8490-9 (Taylor & Francis : alk. paper)
1. Sustainable buildings--Handbooks, manuals, etc. 2. Facility management--Handbooks, manuals, etc. 3. Commercial buildings--Energy efficiency--Handbooks, manuals, etc. I. Title.

  TH880.W66 2009
  658.2--dc22

2008048583

Published by The Fairmont Press, Inc.
700 Indian Trail
Lilburn, GA 30047
tel: 770-925-9388; fax: 770-381-9865
http://www.fairmontpress.com

Distributed by Taylor & Francis Ltd.
6000 Broken Sound Parkway NW, Suite 300
Boca Raton, FL 33487, USA
E-mail: orders@crcpress.com

Distributed by Taylor & Francis Ltd.
23-25 Blades Court
Deodar Road
London SW15 2NU, UK
E-mail: uk.tandf@thomsonpublishingservices.co.uk

Printed in the United States of America
10 9 8 7 6 5 4 3 2 1

0-88173-607-4 (The Fairmont Press, Inc.)
1-4200-8490-9 (Taylor & Francis Ltd.)

While every effort is made to provide dependable information, the publisher, authors, and editors cannot be held responsible for any errors or omissions.

for Lexie…

Table of Contents

Foreword

WHY READ THIS BOOK?

Climate change is a reality that is already affecting organizations (and people) in ways we never imagined. However, beyond minimizing the physical threat of climate change, businesses can prosper TREMENDOUSLY by being leaders and accelerating a new "green" economy. It really comes down to the simple fact that consumers, voters, citizens, and kids want "greener" products and services. As measurement systems improve, "green washing" will not satisfy many of these consumers... *they will want to see real results*. This is good news for facility managers and the reason for this work.

Energy efficiency/conservation is ranked by corporate executives as the #1 way to reduce emissions in a cost-effective manner.[1] Because buildings contribute approximately 43% of the carbon emissions in the US, an opportunity exists to reduce a large part of these emissions and become "environmental heros."[2]

This book's purpose is to showcase some of the PROFITABLE "green" ideas that a business owner can implement QUICKLY. *This book describes PRACTICAL tactics in energy, water, cleaning, transportation, sustainable design, landscaping, recycling, solar, financing, and marketing.* From the board room to the boiler room, you will increase profits and reduce your environmental impact. In addition, you may also "squash" your competition because organizations perceived as more "environmentally friendly" can recruit better faculty, students, suppliers, customers and employees. *Take action now, before your competitors do.* A recent study showed that 92% of young professionals want to work for an organization that is environmentally friendly.[3] Even stock prices of corporations have been proven to improve dramatically when energy management programs are announced.[4] So you are in the right book at the right time!

For decades, the authors of this book have been helping large companies reduce costs via energy-efficiency, waste reduction and productivity improvement programs. Now, beyond saving dollars (which go directly to the bottom line), being "lean" and "green" has proven to attract more sales, which increase profits further.

So why be "Green?"
Because if you don't, *your business may simply not survive.*

Footnotes

1. "Getting Ahead of the Curve: Corporate Strategies That Address Climate Change," Pew Center on Global Climate Change, 2006
2. Pew Center on Global Climate Change, "The U.S. Electric Power Sector and Climate Change Mitigation and Towards a Climate Friendly Built Environment," *2006*.
3. *Wall Street Journal*, November 13th, 2007—"How Going Green Draws Talent, Cuts Costs."
4. Wingender, J. and Woodroof, E., (1997) "When Firms Publicize Energy Management Projects: Their Stock Prices Go Up"—How much—21.33% on Average! *Strategic Planning for Energy and the Environment*, Summer Issue 1997.

Contributors

CHAPTER 1, 13, 14

Eric A. Woodroof, Ph.D., C.E.M., CRM, shows clients how to make more money and simultaneously help the environment. During the past 15 years, he has helped over 250 organizations and governments improve profits with energy-environmental solutions. He has written over 25 professional journal publications and his work has appeared in hundreds of articles. Dr. Woodroof is the chairman of the board for the Carbon Reduction Manager Certification Program and he has been a board member of the Certified Energy Manager Program since 1999. Dr. Woodroof has advised clients such as the U.S. Public Health Service, IBM, Pepsi, Ford, GM, Verizon, Hertz, Visteon, JPMorgan-Chase, universities, airports, utilities, cities and foreign governments.

He is friends with many of the top minds in energy, environment, finance and marketing. He is also a columnist for several industry magazines, a corporate trainer, and a keynote speaker. Eric is the founder of ProfitableGreenSolutions.com and can be reached at 888-563-7221.

CHAPTER 3

Steve Doty, PE, CEM, LEED AP, is an energy engineer for Colorado Springs Utilities, where he provides energy audits for commercial and industrial customers and speaks to business associations. Steve's background includes HVAC design, automatic controls, indoor air quality, and commissioning, all with hands-on project experience. He has written multiple articles for both ASHRAE and AEE, and is a published author and co-editor of two energy handbooks.

Marco Clerx, P.E., C.E.M., C.M.V.P, C.B.C.P., C.E.P.P., D.G.C.P., is the measurement & verification services leader for Honeywell Building Solutions, responsible for ensuring energy saving projects include practical performance measurement and verification (M&V) plans to support the technical and financial guarantees and meet with the client's expectations for savings. Teaming with contract, financial, and project installation managers, he reviews the technical, financial, contractual and relationship concerns potentially leading to future unplanned costs, or operational and/or guarantee performance issues, statutory procurement concerns, and avoidance of possible future interpretation of invalid contract.

Mr. Clerx has over 25 years of energy-related experience at various engineering and utility consulting firms, with an ESCO, with Lawrence Berkeley National Laboratory, and with University of California Riverside. He is a registered professional engineer in four states. He is published in the *Journal of Solar Energy*, and the recipient of the 2007 Region 1 AEE Energy of the Year. He has helped many different kinds of customers including schools, hospitals, municipalities, universities, with projects ranging from wood chip boilers, solar photovoltaics and cogeneration to lighting and domestic water retrofits.

Venkat Iyer, P.E., CEM, LEED AP, has been in the performance contracting business since 1990 working in various vertical markets such as K-12, healthcare, industrial and commercial fields. He has been active in the Honeywell Performance Contracting Engineers group, as a council member (2006-2008) and the technical committee chairman (2007) and council chairman (2008). He has been proactive in standardizing processes in the field such as the development of energy conservation measure binders and data archiving. Venkat has active involvement in the field of renewable energy resources, particularly biomass. The first successful biomass thermal plant in the Arnot Ogden Healthcare Center in Elmira, NY, and several successful solar PV installations in Long Island are testimonies to his accomplishments in the renewable energy field. He is part of a team that has a pending patent application for biomass wood chips field storage and delivery management system. Currently, he is the manager of the performance contracting engineering team for the East Region of Honeywell Energy Retrofit Group.

Venkat Iyer, P.E. , LEED AP

PCE Manager - East Region

Honeywell Energy Services

Phone: (973)-455-3357; Fax: (973)-455-3476

Mobile: (908)-391-4084

101 Columbia Road, Morristown , NJ - 07962

venkat.iyer@honeywell.com

CHAPTER 4

Paul B. Breslow, Ph.D., is the senior renewable energy analyst responsible for developing the core analytic engines and processes for PV system design optimization, production estimation, and financial modeling for the large-scale commercial and utility solar PV integrator, Suntech Energy Solutions, Inc. (formerly Ei Solutions, Inc.). He also supports the sales, marketing, and business development teams in strategy, proposal, project and territory development.

Paul was previously the business development manager for the 3 Phases Energy Services Onsite Division. There, he developed a proprietary solar PV energy output and project finance model for use by the sales and development teams, and initiated key business development efforts. Prior to 3 Phases, Paul cofounded e2d, a renewable energy consulting company focused on solar and wind project and market feasibility.

Paul has a BSE, an MSE, and Ph.D. in mechanical engineering, and MBA coursework from Tulane University in New Orleans, where he focused on the use of weather in electricity demand forecasting and renewable energy modeling.

He was the executive director of the Renewables Council of Louisiana prior to Hurricane Katrina; and he is currently a cofounder of the successor organization, the Louisiana CleanTech Network, which is dedicated to developing the clean technology industry in Louisiana. He is LEED trained and EIT certified. Paul lives in Pasadena, CA, with his wife, Hannah.

CHAPTER 5

Arjun Sarkar has been passionate about alternative fuels for over 20 years. He is a leader within the transportation department at the University of California, Santa Barbara. Arjun is helping develop the alternative fueled vehicle fleet as well as other transportation systems. He has worked with the Santa Barbara Metropolitan Transit District on their electric shuttle program. He is also a key member of the UCSB Sustainability Team and for the past 8 years, he has been the coordinator of the "Green Car Show" at Santa Barbara's world renowned Earth Day Festival. Arjun describes himself as a "Radiant Catalyst for the Renewable Energy Transition."

Arjun's current focus is on helping California meet its alternative fuel targets set forth by the California Energy Commission and the Air Resource Board in AB32, AB1007 & AB118.

Arjun can be reached at: Arjun.Sarkar@tps.ucsb.edu

CHAPTER 6

Bill Jacoby has 23 years of public sector water conservation management, most of which was served in the drought-laden area of Southern California, where water conservation techniques are mature. He has also con-

tributed to the water conservation industry via the following associations:
- California Urban Water Conservation Council person of the year.
- WateReuse Association, CA Section recycling advocate of the year.
- Bemidji State University outstanding alumni of the year.
- Conservation: CUWCC—Convener and chair of various committees, AWWA—Chair of the National Communication Education and Legislation Committee, ACWA—Member of the WUE and Water Management Committees
- Recycling: WateReuse—President of CA Section, chair of CA Legislative/Regulation Committee, member of the National Legislative Committee; ACWA—Member of the Water Recycling Committee; CALFED—Member of the WUE Advisory Committee; and DWR—Member of the State Water Plan Advisory Committee.

CHAPTER 7

Founder of everGREEN landscape architects, **Leland Walmsley** was born into a family of professionally renowned designers practicing architecture, landscape architecture, construction, and city planning in Southern California since 1918.

An accomplished artist and landscape architect in his own right, Leland continues his family's rich tradition of award-winning design, delighting clients, and saving the planet, one gorgeous garden space at a time. "My family and I have always practiced 'cutting edge,' 'organic,' 'environmentally sustainable,' and 'green' design, while others are just now scrambling to catch up. Many of my early clients don't even realize their gardens are 'green,' but are enjoying all the benefits."

In 2003, Leland launched his own firm and is proving a path for others to follow in both awe inspiring and eco-conscious garden design.

In 2006, he received the prestigious Santa Barbara County "Green Award," had a garden featured on the "Tour of Green Homes," and yet another garnered a 6-page spread, "The Privacy Fantasy," in *House Beautiful* magazine.

In 2007, he was sought after to be one of the select few invited to participate in the LA Garden Show, and won the Silver Award at the San Francisco Flower and Garden Show—the 3rd largest show of its kind in the world.

Driven to continuously grow and improve, Leland recently raised the bar by becoming the 1st and only LEED AP* and Green Roof AP landscape architect in Santa Barbara and Ventura Counties.

Leland is often a guest speaker at various local groups of influence, speaking on the subjects of "water efficiency," "getting back to green design," and "greening your home and garden." A founding member of the US Green Building Council regional board, he is also a respected member of Built Green, Green Building Alliance, and several other professional associations.

His firm has adopted the 2030 Challenge—a commitment to landscape design, construction and maintenance with zero fossil fuel emissions by 2030. He even drives a sleek biodiesel powered by soybeans.

Leland has created lush, eco-friendly garden paradises for the founder of Kinkos, the CFO of Oracle, notable celebrities, as well as the private and corporate gardens of two of the largest flower growers on the central coast, just to name a few. "We do it all... from wild habitat restorations, private garden retreats, historic renovations, commercial resort projects, campuses, parks... We even advise our clients on their real estate purchasing options to maximize their understanding of each property's advantages and challenges in order to make the best choice for their needs."

Leland and his talented team of expert professional associates have a true passion for seamlessly blending great garden design, sustainable science, and environmental stewardship.

Leland Walmsley can be reached at : leland@everGREEN-sb.com, or http://www.evergreen-sb.com/

CHAPTER 9

Mary J. Conrad is president of Conrad's Cleaning Service, Inc. in Florida. Conrad's Cleaning Service, Inc. has been in service on the Treasure Coast since 1994. Mary changed from the traditional way of doing business in 2006 to an environmentally friendly manner. Conrad's Cleaning Service, Inc. went green from its home office to the services offered to our clients. Mary actively speaks at local business groups on the Treasure Coast in attempts to convince other businesses how going green is good business. Mary speaks about how going green is good for their business and the environment. Mary is a member of the Building Service Contractors Association International, International Sanitary Supply Association, Treasure Coast Builders Association Green Building Council, and National Association of Women in Construction and the Indian River Chamber of Commerce.

Deborah A. Pate is vice president of Conrad's Cleaning Service, Inc. in Florida. Deborah assists Mary with speaking at local meetings spreading the green message. Deborah is a member of the Building Service Contractors Association International, International Sanitary Supply Association, Treasure Coast Builders Association of Green Building Council, and National Association of Women in Construction.

Deborah's service to her community even included service in the United States Coast Guard from 1992-1997. Deborah admits this is where she spent most of her service time learning to clean government buildings, boats, and grounds to "Uncle Sam's Liking." Deborah admits that times are changing and government and civilian businesses are encouraged to embrace the green movement and move from their traditional way of doing business to be more environmentally friendly.

CHAPTER 11

Ryan Park is one of the most influential individuals in the downstream integration market of the solar electricity industry. He was one of the founding members of REC Solar Inc., which now installs more solar electricity systems every year than any other company in the United States.

Ryan is currently responsible for developing the commercial sales team, structuring large solar projects, and establishing new strategic partnerships. Ryan graduated with honors from California Polytechnic (Cal Poly), San Luis Obispo, and is committed to improving the world through renewable energy technologies, energy efficiency, and empowering people.

His email address is: rpark@recsolar.com

CHAPTER 12

Mike Taylor is the director of research for the Solar Electric Power Association (SEPA), and is responsible for providing technical research, information and support services to electric utility and solar industry members. Prior to joining SEPA in 2006, Mike spent seven years with the Minnesota State Energy Office specializing in renewable energy policy and program development, including the development and operation of Minnesota's solar incentive program. Mike received his master's degree in science, technology and environmental policy from the University of Minnesota, and his bachelor's degree in environmental biology from Saint Mary's University of Minnesota.

CHAPTER 13

John R. Wingender, Jr., Ph.D., is a professor of finance as well as the chairman of the Department of Economics and Finance, College of Business Administration at Creighton University in Omaha, Nebraska. His teaching and research interests are in international finance, investments, and event studies. He has won numerous research and teaching awards and taught at the International Management Center in Budapest, Hungary, in 1993 on a Fullbright Award. His research has been published in many quality business journals such as: *Management Science, The Journal of Financial and Quantitative Analysis, Industrial Relations* and the *Journal of Banking and Finance*. Dr. Wingender has been a consultant for Fortune 500 companies, as well as for private companies. He hopes to work on future interdisciplinary projects.

ABOUT THE SOLAR ELECTRIC POWER ASSOCIATION

From national events to one-on-one counseling, SEPA is the go-to resource for unbiased and actionable solar intelligence. SEPA is comprised of over 250 electric utilities, solar industry companies, and government and research members. Breaking down information overload into business reality, SEPA takes the time and risk out of implementing solar business plans and helps turn new technologies into new opportunities. www.solarelectricpower.org.

PART I

Opportunities to "Green" Your Business

Chapter 1

Easy Ways to "Green" a Business

This chapter is a collection of ideas from discussions/feedback from thousands of people that have attended training classes on energy efficiency and/or carbon reduction strategies with Dr. Woodroof. All of these ideas may not work at every business location, but hopefully you can apply a few to improve your profits and help the planet. The first 3 strategies are "basic" philosophical approaches, while the remaining strategies are more tactical.

1. THINK DIFFERENT AND ELIMINATE
   "HIGH CARBON" PROCESSES

Usually, the best "green" solutions involve "undoing" a solution to another problem that doesn't really exist anymore. For example, why try to minimize the energy consumed by a fax machine, if your business really doesn't need the fax machine anymore (due to scanning and digital technologies)? If you can eliminate the machine, you eliminate the maintenance, supplies and energy required to operate the machine too... and that saves money. Another example might be avoiding the need for lawn pesticides by landscaping with native plants, which also require low amounts of water. Now you have eliminated a lawn mower, the gas required to run that mower as well as the emissions from the mower.

When you really sit down and think about what your business needs today... it is amazing how much "junk" you can eliminate. If you can eliminate that "junk," you may be able to donate it to someone else and get a tax-write off (it may be "treasure" to them). To get started, have a brainstorming session with your employees/team and ask "why" businesses processes exist and keep asking "what is so good about that" until you find a "greener" way to get what you need in today's world. Many Fortune 100 clients have done this and the results are amazing— improved profits and reduced environmental impact.

Someday, people may think of air as an asset that we "own." Every time we generate carbon emissions, we are basically dumping pollution into that asset. Perhaps we should call it "littering," because that is exactly what is happening, we just don't see the litter (emissions) as well as we can see trash on the road. So think about what you are doing and eliminate the big emitters if you can.

If you can eliminate processes that consume an "unfair share" of resources, you will limit your price risk when the cost of those resources escalates. *For example, you probably have changed your car driving habits in the last year due to gas price escalation.* In your organization, to truly get people to eliminate processes (some may view this as a "sacrifice"), we must think different and assign meaningful values to limited resources. *For example, tell someone to "conserve" and it means nothing. However, it might mean more to them if you tell them that for every page we print, it kills part of a tree! Or if we recycle a ton of paper, it saves 17 trees!*

Companies that are "ahead of the game" are encouraging employees to consider the environmental costs that their actions (and company processes) create. As carbon emission legislation is inevitable, proactive companies are already incorporating these social costs *before their competitors*. Additional marketing benefits include claiming emissions savings from eliminating processes or recycling, but we will discuss this later.

Beyond carbon, there are many other ways to save money (and improve morale) by leveraging the famous "3 Rs": Reduce, Reuse and Recycle. To spur your thinking, please take a look at the checklist from a successful green business certification program: http://www.montereybaygreenbusiness.org/. It is a great comprehensive list of green tactics, which can be good discussion points for your team. Go there now and download the checklist; you will get some great ideas!

2.   STOP THE BLEEDING

After eliminating processes, the next best action is to minimize waste/emissions produced from the processes you need. For buildings, a good start is to identify where energy/resources are being used and set up systems to eliminate unnecessary consumption when not needed. This can involve configuring controls on lighting and HVAC systems, training for personnel, etc.. In a manufacturing plant, fixing compressed air, steam or water leaks are good examples that have the quickest returns or "simple paybacks." After turning "off" equipment when not needed, look to energy efficiency technologies for further savings.

3.   MEASURE TO MANAGE

Success is simply defined as getting from point "A" to point "B," so it is dependent on measurement. Although in its "infancy," measurement standards are being developed to define what is "green." Beyond measuring and managing energy, carbon emissions will be a new "key performance indicator" of organizations. *There are talks of having products with labels that include the product's "carbon footprint." It may be very similar to the nutrition information that is on all food products.*

Therefore, to manage and reduce your "footprint" (more accurately stated as an "emissions inventory"), organizations must first begin to measure carbon emissions in a precise way, which is compliant with global standards. Again, the good news is that there are many ways to reduce carbon emissions, which can lead to not only cost-savings measures, but also revenue enhancement! Measuring carbon emission is more complex than measuring energy (due to differences in emission sources from fuel choices), but the savings can be great when you know where to look. It all starts with an "audit" of your carbon emissions. For a free carbon audit, visit: www.freecarbonaudit.com.

If you are a manufacturer and would like a free energy audit, you may qualify for one from the Industrial Assessment Center program (which sends a team of energy specialists to identify energy and waste minimization projects in your facility). You can learn more here at http://www1.eere.energy.gov/industry/bestpractices/iacs.html.

4.   SET UP A FUNDING MECHANISM
      FOR YOUR "GREEN" PROGRAM

Believe it or not, you can get free money to fund your "green" programs/projects. Below are some examples:
- Set up a company credit card that gives cash back towards a "green fund."
- Set up a "green fund" (non-profit organization) and give employees the option to donate to it, just like they currently do for the Red Cross or United Way. No matter what a person's religious/political views

are, we all breathe the same air, and your "green fund" would be used to fund projects that help EVERYONE. You could use this fund to pay for "highly visible" projects (such as solar, wind, hybrids, etc.) on your site or within your organization. *In addition, you could possibly sell the carbon credits from those projects.*

• Utilize an innovative solution/partnership to build revenue without costs. For example, GreenTravelPartners.com is a program that donates the travel commissions back to the organization for use in green projects/programs. You could utilize this program (a partnership with Expedia, Travelocity, etc.) or set up a similar program structure for other services you purchase online.

• As a marketing technique, your company may choose to donate a percentage of revenue back to the environment. The funds collected could flow through your own "green fund" for projects completed on your business site. This approach has helped companies increase sales, while getting many good projects completed!

5.   LEVERAGE UTILITY AND TAX REBATES/CREDITS

Again… it is free money that can be used to fund a project/program. There is a great list of available resources listed by state at: www.dsireusa. org.

6.   LIGHTING PROJECTS

Lighting technology continues to improve every year and the paybacks are some of the quickest in the energy field. Some of the most popular upgrades right now are:

• *Switch from High-intensity Discharge Lamps to High-bay Fluorescents*
• Upgrade to LEDs for signs, specialty lighting and even incandescent in certain applications. *Note on LEDs: Because some LEDs can be used to change the color of a space at different times of the day, they have the ability to enhance the "ambiance" of an environment, which can be used to attract more customers in a retail environment.*

7.   POWER PURCHASE AGREEMENTS FOR
        RENEWABLE PROJECTS

Similar to performance-based contracts, the power purchase

agreement (PPA) allows you to get solar on your roof at no upfront costs. Basically, a third-party pays for the materials and installation and they "sell" you the power generated (from the solar cells on your roof) at rate competitive with your utility. Contracts can vary, but many of the solar projects implemented in California in 2007 and 2008 utilized this unique financing mechanism.

8.   BUILDING COMMISSIONING AND RETRO-COMMISSIONING

"Commissioning" means ensuring that installed systems are operating as designed. In many buildings (even new ones), there are opportunities for improvement as many systems are not operating properly either through incomplete installation, or via a retrofit (where the original design intent may have been lost). As buildings age or organizations change, maintenance practices need to be adjusted and/or systems should be "Retro-commissioned" to avoid unnecessary waste. Studies show that commissioning practices can reduce system operational costs by 5% to 10% or more.

9.   GREEN MARKETING

Not "Green Washing!" Getting credit for your "good" energy/ environmental work is often not done well by engineers. There are some good ways to express the environmental benefits (of your projects) in ways that your CEO (and the public) can appreciate. For example, instead of saying a project will save "X" kilowatt-hours, show that these avoided kWh represent avoided power plant emissions that are equivalent to planting "Y" trees. People can visualize trees, but have a hard time visualizing what a ton of CO_2 emissions looks like. To help you quantify the environmental benefits of your energy-saving projects (barrels of oil not consumed, etc.) you can download a free spreadsheet at profitablegreensolutions.com. A screen shot of the spreadsheet follows.

Alternatively, if you want to demonstrate the environmental impact of your current operations (to get people motivated to change), there is a free carbon audit tool (FreeCarbonAudit.com) which is very helpful.

In addition, if you have one, leverage your marketing department to get your project approved. More often than not, they will be excited about the project you are working on and it helps them develop

PROFITABLE GREEN SOLUTIONS
Carbon Footprint Calculator

INSTRUCTIONS: Type in the kWh and MMBtu (of Natural Gas) you use each year. The calculator will give you a preliminary estimate of your Carbon Footprint (as what it means). *Example numbers are supplied*

Type the amount of Electricity you use *You can find this information on your utility bills.*	750,000 kWh/year
Type the amount of Natural Gas (Methane) you use	50,000 MMBtu/year

Annual Emissions Report

Electricity-Related Emissions (Indirect Emissions from Power Plants):	464 Metric Tons of CO_2
Methane (Natural Gas) Emissions from Stationary Combustion:	2653 Metric Tons of CO_2
TOTAL Emissions (Preliminary- Ignores CO_2e emissions)	**3,117 Metric Tons of CO_2**
Your Total Emissions are Equivalent to:	
Barrels of Oil Being Burned	7,248 Barrels per year
Cars on the Road	571 Cars per year
Gallons of Gas Being Consumed	353,782 Gallons per year
Energy Used by This Many Homes	275 Homes per year
Acres of Pine Trees Being Cut Down	708 Acres per year

Other Sources of Emissions to Consider... Check all that apply:	YES	NO	I Don't Know
Do your organization own or operate vehicles?			
Does your organization buy Propane, Fuel Oil or other Liquid Fuels?			
Does your organization have the service records for your air conditioning repairs?			
Does your organization use refrigerants?			
Does your organization have process emissions?			
Has your organization acquired another that has emissions?			
Does your organization leverage recycling processes for emissions benefits?			
products/suppliers/distributors?			

If you would like a more accurate assessment of your organization's Carbon Footprint (and what you can do to reduce it), please save this form and email to: info@profitablegreensolutions.com Or Call 888-563-7221 and leave a message for Eric Woodroof, Ph.D. Someone will respond to you as soon as possible.

"authentic" green marketing messages for their company, which can improve profits.

10. PARTNER

Due to the rapid pace at which new energy/environmental/carbon technologies are being developed, it is literally impossible for one person (or organization) to know about all of the savings possibilities. It is best to partner with professionals and service companies that are specialists in particular fields such as solar, wind, recycling, chemicals, carbon emissions auditing, etc.. Often, their expertise will add significant value to a program/project and reduce costs as well as time wasted.

CONCLUSION

Obviously, producing a "top ten" list in this rapidly changing field is a little ridiculous. However, CEOs and leaders all over the world are looking for "profitable" solutions. To be direct, THEY ARE LOOKING FOR SOLUTIONS FROM YOU. This is special time for energy/environmental engineers as you can become the CEO's hero!

Chapter 2

Green Janitorial Certification Program

EDITOR'S COMMENTS

A model green certification program is operated by Monterrey Bay (http://www.montereybaygreenbusiness.org/). There is an excellent checklist that covers janitorial, cleaning, and other topics. The guide is free and you should download the latest checklist. However, to show you how comprehensive it is, we have printed the current version below.

GREEN BUSINESS
PROGRAM

Janitorial Companies

Business Name	
Contact	
Phone Number	

MONTEREY BAY AREA GREEN BUSINESS PROGRAM

Janitorial Checklist

Why should my business get certified as a Green Business?

- **No fee**: There is no fee to become a certified Green Business.
- **Better image**: Your company's community image is enhanced through Green Business certification.
- **Save Money**: Saving energy, water and raw materials saves you money. Sending less trash to the landfill saves you money, too.
- **Positive workplace**: Developing a positive, proactive relationship with local compliance inspectors can help you avoid liability, fines and other sanctions.
- **Free advertising**: The Program promotes your business to the public and other businesses for free!
- **Safer Workplace**: Your employees will enjoy a safer workplace and will have one more reason to take pride in working for you.
- **Free assistance**: The Green Business Program offers you free, convenient, time-saving assistance.

How to become a Certified Green Business….

Please read through the following checklist and check all of the boxes that apply. To become certified you must achieve all the criteria that are listed as REQUIRED. There are also some categories that offer choices. Call your Green Business Coordinator if you have questions or need assistance meeting the requirements. There are many rebate programs for facility changes. Before upgrading lighting or water fixtures, Talk to your Coordinator about taking advantage of the free rebates. When you believe you have met most or all of the requirements, contact your Coordinator and he or she will begin the certification process. If there are some things that need to be done to meet the requirements, the Coordinator will let you know what you need to fix before certification.

Remember, the program offers free, non-enforcement, technical assistance to help meet the criteria. We will send out professional technical staff to assist you in meeting the energy, water, resource conservation, and pollution prevention requirements.
- For businesses located in the unincorporated areas of Santa Cruz County, Scotts Valley, and Capitola, call (831) 477-3907 or email: greenbusiness@co.santa-cruz.ca.us
- For businesses located in Monterey County, call Monterey County Environmental Health at (831) 755-4579 or email: NapalanJA@co.monterey.ca.us
- For businesses located in the City of Santa Cruz, call (831)420-5160 or email shealy@ci.santa-cruz.ca.us

Green Business Checklist

Business must meet compliance with regulatory requirements as well as all of the criteria outlined below to obtain Green Business Status, except where a choice is given. If a certain section does not apply to your business, mark it with N/A for Not Applicable. For instance, if there aren't any car washing operations at your facility, mark that section N/A. Offices with a cafeteria that is making and/or serving food will need to meet all Green Business Practices required for restaurants.

A. Pollution Prevention

I. Pollution Prevention Measures and Practices

A. Janitorial Chemicals

(Before GB program)(After GB program)

1. ☐☐ Use Green Seal Certified janitorial products whenever possible.
2. ☐☐ Use non-chlorinated abrasive powders (i.e. Bon Ami)
3. ☐☐ Use a non-chlorine bleach alternative for whitening instead of Hydrogen peroxide or bleach.
4. ☐☐ Use an abrasive sponge or pumice stone in place of strong chemicals to remove grime or deposits.
5. ☐☐ Use a vinegar/water mixture instead of alcohol or ammonia-based window cleaners.
6. ☐☐ Use mild detergents or soaps in place of cleaners with alcohol, ammonia, or caustic ingredients.
7. ☐☐ Screen all products for hazard/toxicity prior to using. Obtain Material Safety Data Sheets (MSDS) for all products used.
8. ☐☐ Be sure to avoid these chemicals:

   - Hydrochloric Acid
   - Phosphoric Acid,
   - Sodium Hydroxide,
   - Sodium Metasilicate,
   - Potassium Hydroxide,
   - Butoxyethanol,

   - Ethanolamine,
   - Toluene,
   - Alkyl Phenol Ethoxylates,
   - Bleach (Sodium Hypochlorite),
   - Paradichlorobenzene (Urinal Blocks),
   - Quaternary Ammonium Chloride (Ammonia)

9. ☐☐ Use the lowest concentration of cleaners that will do the job.
10. ☐☐ Only use zinc-free and butyl-free floor cleaners and strippers. (Go to www.greenseal.org to find a product list.)

B. Cleaning Materials

(Before GB program)(After GB program)

1. ☐☐ Buy cleaning equipment such as vacuum cleaners, mop buckets, mops, that are more durable and energy efficient in order to extend life expectancy and reduce waste.
2. ☐☐ Use biodegradable products, like trash bags, when not cost prohibitive.
3. ☐☐ Use reusable spray bottles for dusters, glass cleaners, etc, instead of disposable aerosols.
4. ☐☐ Whenever possible use spray bottles to apply cleaners, rather than mixing a bucket. (Ensures that less cleaner is used and less is wasted).

C. Drains and Housekeeping

(Before GB program)(After GB program)

1. ☐☐ No wastewater may enter a storm drain. "Only rain down the storm drain.
2. ☐☐ Never hose down or wash floor mats, equipment, or vehicles in an area where the wastewater may flow to a storm drain. Never wash vehicles in an area where the wastewater may flow to a storm drain
3. ☐☐ Use dry cleanup methods as a norm, and sweep prior to mopping floors
4. ☐☐ Dry sweep outdoor seating areas and dispose of the debris in the garbage.
5. ☐☐ Mop water (soapy water only) is discharged to the sanitary sewer, not the storm drain
6. ☐☐ The wastewater from outdoor pressure washing and steam cleaning is routed to the sanitary sewer or to landscaping (not permitted to route to landscaping in the City of Santa Cruz). None of the wastewater is entering a storm drain or neighboring water body.
7. ☐☐ Replace traditional janitorial chemicals, including those used in restrooms and staff break rooms or kitchens, with more environmentally friendly chemicals (i.e. replace Comet with Bonami). Use one or a few multipurpose cleaners, rather than many special-purpose cleaners. If a product is Green Seal Certified, it is typically safer and works well.
8. ☐☐ Correct situations that attract and harbor pests with proper food and garbage storage and landscaping.
9. ☐☐ Use a licensed, registered PCO (pest control operator) for chemical pesticide applications. Only apply pesticides or herbicides during dry weather and never before it rains.
10. ☐☐ Integrated Pest Management - Use (or specify in contracts with landscapers) least toxic pest control methods and products to reduce or eliminate the use of chemical pesticides.

- ¬ Correct situations that attract and harbor pests with proper food and garbage storage and landscaping.
- ¬ Use traps, baits and barriers.
- ¬ Use biological controls.
- ¬ Use pest resistant plants.
- ¬ Use less toxic pesticides such as soaps, oils, and microbials and apply on an "as needed" vs. on a set schedule.
- ¬ When chemical pesticides are necessary, use those labeled "caution" rather than "warning" or "danger"

11. ☐☐ Have a volunteer organization label all storm water drains with "No dumping, Drains to Bay" stencils. Your Green Business Coordinator can organize this for you.
12. ☐☐ Liquids such as leftover beverages are not placed in the garbage, because they eventually reach the dumpster and may leak out into the environment.
13. ☐☐ If water softeners are used, use potassium chloride instead of sodium salt or an exchange service instead of an automatic regenerating unit in areas where treated wastewater is recycled for agricultural purposes.

Compliance Notes

Compliance with environmental regulatory laws is required to be certified as a Green Business. Following are some typical compliance issues that businesses find challenging:

- o No wastewater may enter a storm drain. "Only rain down the storm drain."
- o The wastewater from outdoor pressure washing and steam cleaning is routed to the sanitary sewer or to landscaping (in the City of Santa Cruz this wastewater may not go to landscaping and must go to the sanitary sewer). None of the wastewater is entering a storm drain or neighboring water body. Use a water conserving broom attached to a hose as an alternative to pressure washing where possible.
- o Equipment is not cleaned outdoors where wastewater can enter a storm drain or creek.
- o Mop water (soapy water only) is discharged to the sanitary sewer, not the storm drain.

D. Exterior Storage

(Before GB program)(After GB program)

1. ☐☐ Dumpsters are kept tightly covered and impermeable to rain water. If there are no covers on the dumpster, provide overhead coverage.
2. ☐☐ If the dumpster area has overhead coverage and there is a drain in the area, this drain must be routed to the sanitary sewer or be permanently sealed.
3. ☐☐ Report a leaking dumpster to your waste management agency so it can be repaired or replaced.
4. ☐☐ Post signs at trouble spots (e.g., loading docks, dumpster areas, outside hoses) describing proper practices.
5. ☐☐ Keep receiving and storage areas, parking, landscape, and dumpster area clean and free from litter, oil drips and debris.

Compliance Notes

Compliance with environmental regulatory laws is required to be certified as a Green Business. Following are some typical compliance issues that businesses find challenging:

- o   Dumpsters are maintained leak free. Leaking dumpsters are repaired or replaced immediately.
- o   Dumpsters are kept tightly covered and impermeable to rain water. If there are no covers on the dumpster, provide overhead coverage.
- o   If dumpster areas have overhead coverage and there is a drain in the dumpster area, this drain must be routed to the sanitary sewer. Otherwise, the drain should be permanently sealed.

E. Building and Maintenance Materials and Supplies
Complete four out of items #5-17 below to be certified. Also, please use this section as a reference whenever you remodel.

(Before GB program)(After GB program)

5. ☐☐ Use at least two alternative building/maintenance materials or supplies.
6. ☐☐ Use natural or low emissions building materials, carpets or furniture.
7. ☐☐ Use electric (not gas) powered tools.
8. ☐☐ Use wet scraping, tenting or HEPA-vac instruments to reduce dust and debris when removing paint (avoiding chemical paint stripping).

GREEN NOTES - Green Building

When remodeling your business review the following materials prior to construction:

¬ *City of Santa Cruz Green Building Policy:*
    http://www.ci.santa-cruz.ca.us/pl/gbwg/gbwg.html

¬ Build-it Green: http://www.builditgreen.org

¬ *US Green Building Council:* http://www.usgbc.org/

¬ Santa Cruz Green Building Directory:
    http://www.ecoact.org/PDF/GBD_2007_Application.pdf

---

9. ☐☐ Use high-efficiency paint spray equipment.
10. ☐☐ Buy rechargeable batteries and appliances such as hand-held vacuum cleaners and flashlights.
11. ☐☐ Print promotional materials with soy or other low-VOC inks.
12. ☐☐ Use unbleached and/or chlorine-free paper products (copy paper, paper towels, coffee filters. etc.).
13. ☐☐ Switch from commercial air fresheners to potpourri or vinegar & lemon juice.
14. ☐☐ Switch from toxic permanent ink markers/pens to water-based markers.
15. ☐☐ Purchase laundry detergents that have little or no phosphates.
16. ☐☐ Purchase recycled content construction materials when building/remodeling (such as plastic lumber for decking, benches and railing, carpet, carpet padding, etc).
17. ☐☐ Buy low-mercury fluorescent lamps.

Janitorial Companies

F. Air Emission Reductions

Please complete this requirement:

(Before GB program)(After GB program)

1. ☐ ☐ Encourage employees to use alternative transportation to get to work such as bike, bus, or carpool. Please describe how:

_____

_____

GREEN NOTES – Vehicle Emissions

An improperly tuned car produces 10-15 times more pollution than a tuned one. Each person driving alone to work creates more than 2 tons of auto exhaust each year. If every commuter car in the U.S. carried just one more passenger, we would save 600,000 gallons of gasoline and reduce air emissions by 12 million pounds of carbon dioxide every day!

The amount of carbon emitted for driving 50 miles is roughly equal to 105 square feet of mature forest.

In addition, reduce air pollution in at least two ways:

(Before GB program)(After GB program)

1. ☐ ☐ Encourage commute alternatives by informing employees and customers about transportation options for reaching your location (i.e. post transit schedules/routes).
2. ☐ ☐ Help employees rideshare by posting commuter ride sign-up sheets, employee home zip code map, etc. Get assistance from www.rides.org or 1-800-755-POOL.
3. ☐ ☐ Offer telecommuting and/or flexible schedules so workers can avoid heavy traffic commutes.
4. ☐ ☐ When possible, arrange for a single vendor who makes deliveries for several items.
5. ☐ ☐ Patronize services close to your business (e.g., food/catering, copy center, etc.) and encourage employees to do the same.
6. ☐ ☐ Purchase Carbon Offsets to compensate for miles traveled by company vehicles.
7. ☐ ☐ Other _____

If your business has a company-owned vehicle(s), complete at least two of the following or purchase low emission vehicles or alternative fuel vehicles:

(Before GB program)(After GB program)

1. ☐ ☐ Plan delivery routes to eliminate unnecessary trips.
2. ☐ ☐ Keep vehicles well maintained to prevent leaks and minimize emissions, and encourage employees to do the same.

If you have more than 100 employees, complete at least 3 of the following:

(Before GB program)(After GB program)

1. ☐ ☐ Provide car/van pool parking.
2. ☐ ☐ Provide a commuter van.
3. ☐ ☐ Sell bus or light rail passes on-site or at a discount to your employees.
4. ☐ ☐ Offer a shuttle service to and from bus, train and/or light rail stops.
5. ☐ ☐ Provide shower facilities for employees who walk/jog/bike to work or contract with an athletic club to use their facilities.
6. ☐ ☐ Encourage bicycling to work by offering rebates on bicycles bought for commuting.
7. ☐ ☐ Provide secured and enclosed bicycle parking for employees (e.g., bike lockers)

B. Energy Conservation

I. Energy Management

Please complete the mandatory measures listed below:

(Before GB program)(After GB program)

1. ☐☐ Organize an energy service to conduct a commercial energy audit of your facility to help identify which energy conservation measures to use at your business. Your Green Business Coordinator can organize this for you.

2. ☐☐ Complete regularly scheduled maintenance on your HVAC (heating, ventilation and air conditioning) system.
   - ¬ Clean permanent filters with mild detergents every two months (change replaceable filters every 2 months).
   - ¬ Check entire system each year for coolant and air leaks, clogs, and obstructions of air intake and vents.
   - ¬ Keep condense coils free of dust & lint.

3. ☐☐ If more than *five* people are employed, track and post monthly gas and electricity usage information for employees to view.

II. Energy Conservation Measures and Practices

Implement at least *ten* of the following measures listed below, with at least *four* in Section A:
A. Equipment/Facility Changes:

(Before GB program)(After GB program)

1. ☐☐ Use an energy management system to control lighting, kitchen exhaust, refrigeration and HVAC.
2. ☐☐ Install occupancy sensors for lighting in low occupancy areas, including walk-in refrigerator/freezers.
3. ☐☐ Retrofit incandescent bulbs with compact fluorescent lights.
4. ☐☐ Install ultra efficient ballasts such as GE UltraMAx units to dim lights to take advantage of daylight.
5. ☐☐ Upgrade existing fluorescent lighting with T-8 lamps with electronic ballasts (T-8 systems consume up to 40% less energy than conventional T-12 systems). Recycle old lamps and ballasts.
6. ☐☐ Install a programmable thermostat to control heating and air conditioning.
7. ☐☐ Insulate all major hot water pipes.
8. ☐☐ Insulate refrigeration cold suction lines.
9. ☐☐ Use weather stripping to close air gaps around doors and windows.
10. ☐☐ Retrofit exit signs with LEDs or fluorescent bulbs.
11. ☐☐ Select electrical equipment with energy saving features (e.g. Energy Star).
12. ☐☐ Install and use computer hardware programs that save energy by automatically turning off idle monitors and printers.
13. ☐☐ Plant native shrubs or trees near windows for shade.
14. ☐☐ Install plastic strip curtains on walk-in refrigerator/freezer doors.
15. ☐☐ Convert hot water heaters to on-demand systems.
16. ☐☐ Use a solar water heater or pre-heater.

17. ☐☐ Reduce the number of lamps and increase lighting efficiency by installing optical reflectors or diffusers.
18. ☐☐ Install ceiling fans.
19. ☐☐ Consider adding desk lamps or task lighting fixtures to work spaces in order to reduce the need for overhead lighting when only one person is in the office.
20. ☐☐ Other _____

B. Employee Practices

(Before GB program)(After GB program)

1. ☐☐ Clean lighting fixtures and lamps so that they are lighting as effectively as possible (dirt can reduce lighting efficiency by up to 50%) and replace aging fluorescent tubes. Then remove lamps where possible.
2. ☐☐ Check and adjust lighting control devices such as time clocks and photocells.
3. ☐☐ Set thermostat to 76° F for cooling, 68°F for heating, and use the thermostat's night setback.
4. ☐☐ Institute a policy that all electronic devices and lighting be turned off in non-occupied rooms.
5. ☐☐ Drain and flush hot water tanks to the sanitary sewer every 6 months to prevent scale build up and deposits (This can reduce heating efficiency).
6. ☐☐ Set hot water heaters to standard 140-150° F.
7. ☐☐ Turn room cooling units off when the weather is cooler.
8. ☐☐ Maintain refrigerator doors by replacing worn gaskets, aligning doors, enabling automatic door closers, and replacing worn or damaged strip curtains.
9. ☐☐ Maintain refrigerators by keeping evaporator coils free of excessive frost and by keeping condenser coils free of dust and lint.
10. ☐☐ Check pilot lights for proper adjustment
11. ☐☐ Rearrange workspace to take advantage of natural sunlight, and design for increased natural lighting when remodeling.
12. ☐☐ Use light switch reminders to remind customers and staff to turn off lights.
13. ☐☐ Use energy efficient space heaters during off hours instead of heating the whole office.
14. ☐☐ Other _____

> **GREEN NOTES – Office Electronics**
>
> *Energy Star® compliant monitors have power management features and consume up to 90% less energy. Screen savers don't save energy! Energy Star® copiers and fax machines can reduce their annual electricity costs by about 60% and 50% respectively.*

C. Solid Waste Reduction

I. Waste Management

Have a solid waste reduction assessment done for your facility to help identify which waste reduction, reuse and recycling practices would best work for your shop. Your Green Business Coordinator can organize this for you. **Please complete the mandatory solid waste measures listed below:**

(Before GB program)(After GB program)

1. ☐☐ Eliminate the use of polystyrene containers. Utilize one of the following options (in order of preference): paper, paperboard, compostable containers (starch-based sugarcane, rice hulls, and/or corn), or recyclable plastic.
2. ☐☐ Where applicable, provide recycling and composting container(s) at convenient and appropriate locations.
3. ☐☐ If you provide disposable bags to your customers for items purchased or supplied by your business, you must primarily provide paper bags instead of plastic. Plastic bags should only be provided when necessary.

> **GREEN NOTES - Recyclables**
>
> *Santa Cruz Regional Recycling Programs continually upgrade recycling capabilities. For detailed information on recycling programs refer to Ecology Action's Recycling Guide:*
>
> *http://www.ecoact.org/PDF/WTR_res.pdf*

II. Waste Reduction Measures and Practices

Perform all *five* of the following activities to reduce paper usage and increase reuse.

(Before GB program)(After GB program)

1. ☐☐ Keep a stack of previously used paper near printers; use it for drafts or internal memos, or designate a draft tray on printers with multiple trays.
2. ☐☐ Purchase/lease all new copiers and printers with double sided copying capability.
3. ☐☐ Encourage employees (signs or memos) to use reusable lunch bags and containers, and for leftovers from restaurant lunches.
4. ☐☐ Set all multi-page documents and defaults on printers and copiers to be double sided.
5. ☐☐ Set document defaults to smaller fonts and margins.

> **GREEN NOTES - Polystyrene**
>
> *Paper takes 4 times less space in storage and disposal than polystyrene. In 1986, EPA ranked the 20 chemicals whose production generated the most hazardous wasted. Polystyrene was number five.*
>
> *In most parts of the county including Santa Cruz, Polystyrene is NOT recyclable regardless of the stamp on the container.*

A. Paper Reduction

Reduce paper in at least *five* of the following ways:

(Before GB program)(After GB program)

1. ☐☐ Use computer fax modems that allow faxing directly from computers without printing or use email rather than faxing.
2. ☐☐ Manage incoming and outgoing mail by completing any of the following:
   ¬ Eliminate unwanted mailings by calling sender's 800 number or writing "refused" on first class mail.

> **GREEN NOTES – Paper Wastes**
>
> *The average office worker discards more than 175 pounds of high-grade office paper each year.*

      ¬ Eliminate duplicate mailings & subscriptions by returning labels to the sender requesting that all but one be removed.
      ¬ For bulk mail, request removal of name, & write "refused" on first class mail.
      ¬ Purge your own mailing lists to eliminate duplication.
      ¬ Re-use envelopes you've received by covering up the old address and postage, and affix new.

3. ☐☐ Set up a bulletin board or develop routing lists for bulletins, memos, trade journals to minimize the number of employees receiving individual copies.
4. ☐☐ Replace memos with e-mail messages & discourage the printing of messages.
5. ☐☐ Design marketing materials that require no envelope – simply fold and mail.
6. ☐☐ Request that marketing materials be printed on recycled content paper.
7. ☐☐ Use electronic billing methods to invoice customers and receive payment.
8. ☐☐ Other_____

B. Other Waste Reduction

Recycle or reuse materials in 2 additional ways:

(Before GB program)(After GB program)

1. ☐☐ Use laundry service that provides reusable bags for dirty and clean linen.
2. ☐☐ Leave grass clipping on mowed turf ("grass-cycling") rather than disposing.
3. ☐☐ Compost food scraps.
4. ☐☐ Compost or recycle landscape debris.
5. ☐☐ For shipping non-food items, use shredded paper for packaging needs instead of purchasing polystyrene pellets, bubble wrap, other packing materials (if you receive these, reuse them in your own packaging).
6. ☐☐ Printer cartridges.
7. ☐☐ Plastic wrap/bags.
8. ☐☐ CD's/DVD's (mail to: Envirom, 22605 E. La Palma Ave., Suite 501, Yorba Linda, CA 92887).
9. ☐☐ Other _____

> **GREEN NOTES – Regional Landfill Status**
>
> ※ A May 2007 study by a local environmental task-force slated the Buena Vista Landfill to reach capacity in less than fifteen years.
>
> ※ Board of Supervisors approved diversions goals including:
> - Diversion rate of 75% by 2010
> - Recycling programs for building materials and food wastes

III. Environmentally Preferable Purchasing

Please complete the mandatory environmentally preferable purchasing measures listed below:

(Before GB program)(After GB program)

1. ☐☐ Purchase 100% recycled content office paper with at least 30% being post consumer waste (pcw).
2. ☐☐ Purchase paper towels for restrooms with the highest recycled content available.

A. Purchasing Activities

Demonstrate a commitment to purchasing Environmentally Preferable Products. **Select at least *three* of items 1 through 10 to reduce the environmental impact of purchasing decisions.**

(Before GB program)(After GB program)

Janitorial Companies

1. ☐☐ Centralize purchasing to eliminate unnecessary purchases and ensure that all waste reduction purchasing policies are followed.
2. ☐☐ Select products shipped with less packaging or that are shipped in returnable, reusable or recyclable containers.
3. ☐☐ Work with vendors to minimize product packaging: ask vendors to take back packaging and used/damaged products for reuse and recycling (choose vendors that offer these services).
4. ☐☐ Arrange for cooperative buying through association, co-located business groups, etc.
5. ☐☐ Purchase reusable rather than disposable office items such as refillable pens, erasable white boards & wall calendars.
6. ☐☐ Have all employees use reusable mugs and cups. In the lunch/break room, eliminate disposables by using permanent ware (mugs, dishes, utensils, towels, rags, coffee filters, etc.) and using refillable containers of sugar, salt & pepper, etc. to avoid individual condiment packets.
7. ☐☐ Donate old uniforms and linens to shelters or nonprofits or otherwise recycle them.
8. ☐☐ Buy products that are bulk, concentrated, durable, repairable, and/or recyclable, making sure that you need ALL you are ordering.
9. ☐☐ Retailers – stock/sell products which are less toxic or less polluting than conventional products.
10. ☐☐ Retailers - offer an incentive to customers who bring their own shopping bags, coffee mugs, etc.

B. Purchases

Purchase *three* recycled content products from items listed below:

(Before GB program)(After GB program)

11. ☐☐ Business cards.
12. ☐☐ Disposable drink and food items.
13. ☐☐ Storage bins and containers for recyclables.
14. ☐☐ Refuse pails and bags (recycled HDPE trash liner bags instead of LDPE or LLDPE).
15. ☐☐ Floor mats.
16. ☐☐ Toilet seat covers and toilet paper.
17. ☐☐ Carpet.
18. ☐☐ Construction materials when building or remodeling.
19. ☐☐ Use recycled-content paint.
20. ☐☐ Pencils, rulers and other desk accessories.
21. ☐☐ Purchase mulch, soil amendments and compost made of plant trimmings or green waste.
22. ☐☐ Other:_____

> **GREEN NOTES – Office Paper**
>
> *In the manufacture of "recycled" paper, 64% less energy and 58% less water is required, and 74% less air pollution is generated.*
>
> *Look for recycled paper with a high post-consumer content (previously used-not manufacturing scraps). Copy paper with 30% post-consumer content is readily available and proven effective.*

Recyclables and Hazardous Wastes

The following items are recyclable either by curbside collection or at a drop-off site depending upon your refuse service and location:

- aluminum
- bottles
- cans
- cardboard
- glass
- magazines and junk mail (remove any plastic)
- metals
- mixed paper (except bright or fluorescent paper)
- newspaper
- office paper
- packaging
- paint (if water-based)
- phone books
- plastic bags (must be placed in a larger plastic bag that is tied off)
- plastic containers (# 1-7 on container) except black plastic
- plastic sheets (no tarps)
- printer cartridges (recycle or refill at local participating stores, your Coordinator can inform you)
- tires
- wood pallets and spools
- yard waste

Recyclable Hazardous Wastes

The following items are hazardous waste and recyclable if taken to a Household Hazardous Waste collection facility (where available) or picked up by a hazardous waste hauler:

- automotive fluids, i.e. coolant and waste motor oil
- electronic equipment (anything with a cord)
- oil-based paints and solvents
- fluorescent light bulbs (tube and CFL)

**Note: If your business is a small quantity generator (SQG) of hazardous waste, then you may recycle hazardous wastes at a Household Hazardous Waste facility. If your business is a large quantity generator, you must have a licensed hazardous waste hauler remove and recycle these waste streams. If your business generates less than 1000 kilograms (2200 pounds), then you fall under the SQG category.

Electronic Wastes

Electronic wastes such as computers, televisions, printers, cell phones, etc. can be taken to landfill household hazardous waste facilities, provide your business is a SQG. However, the best option for this wastestream is reuse through our Community Partners, where electronics can be reconditioned for the public to benefit the community. Please visit http://www.ecoact.org/Programs/Waste_Reduction/Recycling/ewaste.htm for details on reuse of e-waste.

Janitorial Companies

D. Water Conservation

I. Water Management

Contact your water utility to request a free water use survey of your facility (where available) and ask for your available water usage data-preferably for the past three years. You should also ask about their rebate programs. (Your Green Business Program Coordinator can organize this for you). Review the water use survey results annually to identify additional ways to reduce your water use and retain all your future water use data.

Complete all of these mandatory water conservation measures listed below that are applicable to your business:

(Before GB program)(After GB program)

1. ☐☐ Understand your water bill and review it monthly for indications of leaks, spikes or other problems. Call your water utility if you notice any unusual increases in use or if you need suggestions on how to improve the efficiency of your water use.
2. ☐☐ Learn how to read your water meter.
3. ☐☐ Regularly check for and repair all leaks in your facility (toilet leaks can be detected in tank toilets with leak detecting tablets, which may be available from your local water company). Train your staff to monitor and respond immediately to leaking equipment.

> **GREEN NOTES**
>
> *A faucet with a slow leak can waste 10 gallons of water a day, or more! A single leaky toilet can waste as much as 1000 gallons of water per day.*

4. ☐☐ Use "dry sweeping" to clean concrete or asphalt surfaces instead of using water to wash down surfaces. Use high pressure, low water use cleaning techniques only when necessary. Always send wastewater from pressure washing to landscaping or the sewer (discharge to the sewer is mandatory in the City of Santa Cruz), not the storm drain. Use a water conserving broom attached to a hose as an alternative to pressure washing where possible.

II. General Water Conservation Measures and Practices

Complete the mandatory water conservation measure directly below if it is applicable:

(Before GB program)(After GB program)

1. ☐☐ Install low flow aerators in faucets (1.5 gpm) and showerheads (2.5 gpm). Your water utility may provide these for free.
2. ☐☐ Install toilets manufactured to flush 1.6 gallons or less. Your water utility may have a rebate program for low flow toilets.

Implement at least 3 of the elective water conservation measures listed below. Consider areas of greatest water use at your facility in choosing new measures.

A. Fixtures and Equipment

(Before GB program)(After GB program)

1. ☐☐ Install non-water (water free) urinals.
2. ☐☐ Install urinals that are manufactured to flush at 1.0 gallon or less. Or replace diaphragms in the flush valve, so that they flush 1.0 gallon.

3. ☐☐ Install low flow, self-closing faucets either infrared or spring–loaded.
4. ☐☐ If cleaning floors with water, use high-pressure low-volume cleaning equipment or use a recycling filtered system such as, an electronic powered cleaning machine.
5. ☐☐ Maintain water pressure (pressure reducing valve) between 60-80 PSI to optimize performance and reduce water loss through leakage, if necessary.

B. Indoor Water Management Practices

(Before GB program)(After GB program)
1. ☐☐ Change window-cleaning schedule from "periodic" to "as required."
2. ☐☐ Use dry floor cleaning methods indoors followed by damp mopping, rather than spraying or hosing with water.

C. Outdoor Water Management Practices

(Before GB program)(After GB program)
1. ☐☐ Instead of washing vehicles on site, go to a washing service that recycles water.
2. ☐☐ Regular pavement cleaning is accomplished by sweeping manually or with electric vacuum or blower, and properly disposing of debris.

D. Other (describe):

_____

_____

_____

III. Landscaping

Complete all of these mandatory water conservation measures that are applicable to your business:
(Before GB program)(After GB program)
1. ☐☐ Test irrigation sprinklers 4 times per year to ensure proper operation and coverage.
2. ☐☐ Repair all broken or defective sprinkler heads/nozzles, lines & valves.
3. ☐☐ Adjust sprinklers for proper coverage – optimizing spacing and avoiding runoff onto paved surfaces. Adjust sprinklers to achieve even water distribution.
4. ☐☐ Adjust sprinkler times and/or duration according to seasons, water during non-daylight hours (generally before 7 am or after 9 pm).

If you have landscaping, you must meet at least 3 of the elective water conservation criteria below:

(Before GB program)(After GB program)
5. ☐☐ Rain shut-off devices or moisture sensors are installed to override automatic irrigation when adequate moisture exists.
6. ☐☐ The number of days lawns are irrigated is limited to a maximum of 3-4 days per week during summer, 2-3 in the spring and fall, and none in the winter. Tree and shrub watering is limited to a maximum of 2 days per week in the summer, 1-2 days in the spring and fall, and none in the winter. Foggy coastal areas can usually get by with fewer days per week of irrigation in the summer season.
7. ☐☐ Prevent runoff when irrigating landscaping on slopes or in narrow planting strips, by scheduling multiple run times for short periods (3-5 minutes), with at least an hour between water applications.
8. ☐☐ Valves are separated based on plant water use (hydro zones).
9. ☐☐ Sprinklers are matched with same precipitation rates.

10. ☐☐ Automatic irrigation controller has the following features:
  ¬ Dual programming capability program A and B
  ¬ Automatic rain shut-off
  ¬ Soil moisture sensor to override program when adequate moisture is present

11. ☐☐ At least two inches of mulch is applied in all non-turf planting areas.
12. ☐☐ Plant material is native or drought tolerant (water conserving).
13. ☐☐ Where available, use recycled water instead of potable water for landscaping.
14. ☐☐ Demonstrate/describe your alternative water conservation techniques for
  landscaping:_____

  _____

  _____

Janitorial Companies

E. Employee Awareness

1. ☐ New and current employees are trained to follow the Green Business practices.
2. ☐ All employees are trained on hair care products and health hazards associated with them
3. ☐ An employee will be asked if they know what Green Business and/or Best Environmental Practices are and they will be asked to give an example.
4. ☐ Provide incentives to employees who take ownership of Best Environmental Practices such as "Employee of the Month".

F. Compliance Checks

1. ☐ Business has not had any SIGNIFICANT health violations that have not been corrected (confirm with Environmental Health Services/Consumer Protection Agency)
2. ☐ Business has met compliance with all storm water-related regulatory requirements (confirm with Environmental Health Services/Certified Unified Program Agency and regional Publicly Owned Treatment Works [POTW])
3. ☐ Business has met compliance with all wastewater-related regulatory requirements (confirm with regional POTW Pretreatment Programs)

G. Client Recommendations

Now that your business is aware of how to prevent pollution, recycle and conserve energy and water, make recommendations to your clients where appropriate:

1. ☐ Look for leaky toilets and faucets, point them out to your clients and explain how fixing the leaks can save your client money on both the water bill and the sewer bill.
2. ☐ Look for old appliances that are inefficient, explain rebate programs for energy efficient appliance purchases.
3. ☐ Help your client set up a recycling system that is easy and works best for them. Explain what is and isn't recyclable.
4. ☐ Talk about your environmentally-friendly cleaning process and educate your clients on why it is better.

All criteria have been met as of the following date: _____

Signature of authorized Green Business Program Coordinator:

_____

Printed Name:

Chapter 3

Energy Conservation And Audits

EDITOR'S COMMENTS

Most people know that energy efficiency or conservation will reduce environmental impact. There are literally millions of ways to be more energy efficient and there are not enough pages in a forest to print them, but this chapter has three parts: A, B and C. Part A is a list of the typical energy conservation measures you can find by facility type. Part B is a more detailed description of some of the energy conservation measures. Part C is a mini-chapter describing the energy audit process and tools. You can also find other books on energy auditing as well as examples to help you further. This is a just a short list of good ideas and methods.

PART A

Energy Saving Opportunities by Business Type

HOW TO READ THIS CHAPTER

The dominant energy users are listed first, which are the drivers of utility costs and the most likely categories to bring the largest results. There will be other measures that can be suggested at each facility.

Common measures are organized in groups, which are:

- Controls
- Maintenance
- Low-cost/No Cost
- Retrofit—Or Upgrade at Normal Replacement
- May Only Be Viable During New Construction

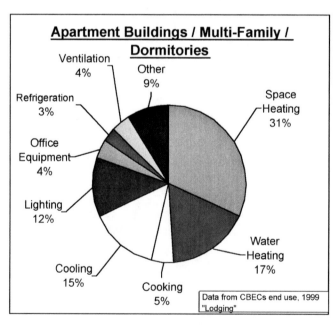

Figure 3A-1. Energy End Use Distribution for Apartment Buildings/ Mullti-family/Dormitories

APARTMENT BUILDINGS/MULTI-FAMILY/ DORMITORIES

Primary Energy Use Sources
- Heating and cooling
- Water heating
- Lighting

Controls
- Controls to lock out cooling below 50 deg F.
- Controls to lock out heating above 65 deg F.
- Reset boiler (HWS) based on outside air temp.
- Lower domestic hot water temperature to 120 deg F.
- Turn off domestic hot water re-circulation pumps at night.
- Occupancy sensors in amenity areas (fitness room, etc.).

Maintenance
- Calibrate controls every two years.
- Annual maintenance on all heat transfer surfaces, including good quality filters, cleaning coils, cleaning tubes, cleaning apartment refrigerator coils.
- Look for open windows in cold weather as signs of defective controls.
- Repair or replace defective zone control valves.

Low Cost/No Cost
- Thermostats with a minimum of 5 degrees deadband between heating and cooling.
- Low flow faucets and shower heads.
- Insulate bare hot water piping.
- Flyers to encourage energy conservation. Possible shared savings or picnic, etc.

Retrofit—Or Upgrade at Normal Replacement
Heating and Cooling
- Higher efficiency heating and cooling equipment.
- Increase roof insulation thickness to current energy code level as part of roof replacement.
- Window replacement.

Water Heating
- Condensing domestic water heater.
- Stack damper for gas-fired domestic water heater.

Lighting
- Higher efficiency lighting in common areas.
- Higher efficiency lighting in room areas, esp. to replace incandescent lighting.

Other
- Low water volume/high speed spin wash machines (E-Star).
- Energy Star Appliances on replacement.

May Only Be Viable During New Construction
- Envelope construction to match local residential energy code for insulation.
- Overhangs or other exterior shading at windows.
- Toilet exhaust on demand (light switch) instead of continuous.
- Low volume bathtubs.
- High performance glass on all windows.

Pool: See "Pools" in Part A.
If a boiler is used: See "Boilers" in Part A.
If a chiller is used: See "Chillers" in Part A.
If the building is over four stories in height: See "High Rise" in Part A.

CHURCHES/WORSHIP

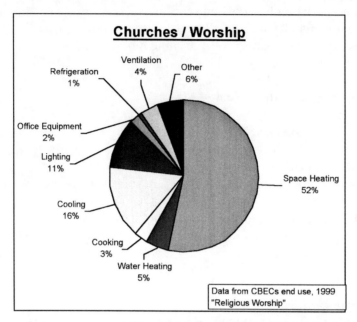

Figure 3A-2. Energy End Use Distribution for Churches / Worship

Primary Energy Use Sources
- Heating and cooling
- Lighting

Controls
- Scheduled start-stop of HVAC equipment and lighting.
- Space temperatures set for 75 cooling/70 heating during services, and allowed to float higher and lower in other times, and at night.
- Morning warm-up or cool-down before services with outside air closed.
- Demand controlled ventilation for variable occupancy.
- For all roof openings to hoods and equipment not active in winter, dampers should be tightly closed during heating operation and when roof equipment is off.

Maintenance
- Calibrate controls every two years.
- Annual maintenance on all heat transfer surfaces, including good quality filters, cleaning coils, cleaning tubes.

Low Cost/No Cost
- Programmable thermostat.

Retrofit—Or Upgrade At Normal Replacement
Heating and Cooling
- Higher efficiency heating and cooling equipment.
- Anti-stratification fans for high bay, heated areas to move warm air to the floor.
- Reduce outside air to proper quantities if excessive.
- Increase roof insulation thickness to current energy code level as part of roof replacement.

Lighting
- Higher efficiency lighting
- Replace incandescent lighting with more efficient technology.

May Only Be Viable During New Construction:
- Passive shading elements for large glass areas.
- High performance low E glazing
- Exhaust heat recovery for high ventilation times.

If a boiler is used: See "Boilers" in Part A.

DATA CENTERS

Primary Energy Use Sources
- Computer equipment
- Cooling
- Lighting (distant third)

Lighting: Higher efficiency lighting and occupancy sensor for unoccupied times.

Table 3A-1. Computer Data Center Energy Savings Opportunities
Source: "Energy Efficiency in Computer Data Centers," Doty, *Energy Engineering Journal*, Vol. 103, No. 5, 2006.

System Type	Measure	New or Retrofit	Basis of Savings
Chilled Water	Size Coils for 50 degF CHW, and Use Elevated CHW temp.	NEW or RETROFIT	Prevent simultaneous dehumidification / humidification
	Variable Speed Fans	NEW	Reduce parasitic fan heat losses.
Air-Cooled DX	Generous Sizing of Air-Cooled Condensers	NEW	Improve heat transfer and reduce approach temperature, improved refrigeration cycle.
	Keep Condenser Surfaces Clean	NEW or RETROFIT	Reduce head pressure for improved refrigeration cycle.
	Adjust Head Pressure Regulation Devices	NEW or RETROFIT	Prevent unintended false-loading of the refrigeration equipment in warm weather.
	Maintain Outdoor Equipment Spacing.	NEW	Prevent air recirculation, keep head pressure low for improved refrigeration cycle.
	Evaporative Pre-Cooling	NEW or RETROFIT	Reduce head pressure for improved refrigeration cycle. Reduce demand.
Water-Cooled DX	Select Equipment to Operate at Reduced Condenser Water Temp.	NEW	Reduce head pressure for improved refrigeration cycle.
	Use a Fluid Cooler instead of a Dry Cooler	NEW or RETROFIT	Reduce head pressure for improved refrigeration cycle.
	Adjust Water Regulation Valves	NEW or RETROFIT	Reduce head pressure for improved refrigeration cycle.
	Use Auxiliary Cooling Coils with a Fluid Cooler.	NEW or RETROFIT	Evaporative cooling reduces load on the refrigeration system, by pre-cooling or allowing the compressors to stop.
	Use Auxiliary Cooling Coils and Link to the Central Chilled Water System.	NEW or RETROFIT	Higher efficiency kW/ton at central cooling system compared to computer cooling equipment saves energy and

Reduced run time of computer cooling compressors extends equipment life. |
| Common Items | Raise the Room Temperature | NEW or RETROFIT | Reduce thermodynamic lift by raising the refrigeration cycle low pressure. |

Envelope: Locate away from envelope effects, especially glazing.

Maintenance: Calibrate every two years.

If a chiller is used: See "Chiller" in Part A.

- Water-side economizer

Sub-meter: If combined with other building uses

Table 3A-1 (*Continued*). Computer Data Center Energy Savings Opportunities

System Type	Measure	New or Retrofit	Basis of Savings
Common Items Cont'd	Reduce Overhead Lighting Power.	NEW or RETROFIT	Reduce cooling heat load from lighting.
	Don't Provide Heaters in the Cooling Units	NEW	Savings in electrical infrastructure, including generator.
	Lower room humidity setting to 30% if possible.	NEW or RETROFIT	Prevent simultaneous dehumidification / humidification
	Use evaporative or ultrasonic humidification instead of infrared or resistance heat.	NEW or RETROFIT	More efficient technology for humidifying Part of cooling load displaced by adiabatic cooling
	Oversize Filters	NEW or RETROFIT	Reduce air flow resistance, in turn reducing fan motor kW and parasitic fan heat loss if fan speed is adjusted down.
	Premium Efficiency Fan Motors	NEW or RETROFIT	Reduced fan motor kW and parasitic fan heat loss.
	Raised Floor System Insulation and Air / Vapor Barrier	NEW	Reduced thermal loss to adjacent floor below. Reduced air and moisture losses by proper sealing of the plenum floor and walls, as well as plumbing and pipe penetrations.

EDUCATION—COLLEGES AND UNIVERSITIES

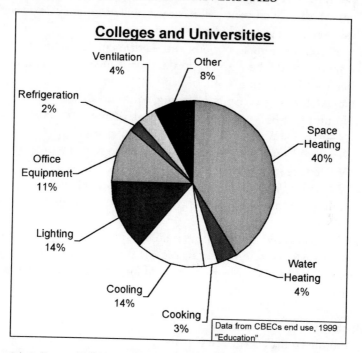

Figure 3A-3. Energy End Use Distribution for Education/Colleges/Universities

Primary Energy Use Sources
- Heating and cooling
- Lighting
- Office equipment

Controls
- Scheduled start-stop of HVAC equipment and lighting.
- Set space temperatures to 75 cooling/70 heating.
- 5 degree deadband between heating and cooling operations.
- Eliminate simultaneous heating and cooling.
- Set space temperatures during unoccupied times up to 85/set back to 60 at night.
- Lock out cooling below 55 degrees and use only economizer.
- Extend the economizer cooling function, from 55 deg F to 65 deg F, if outdoor dew point levels are at or below indoor dew point levels— i.e. if not too humid outdoors.

- Optimum start and stop of primary cooling and heating equipment.
- Demand controlled ventilation for variable occupancy areas such as classrooms and lecture halls.
- For all roof openings to hoods and equipment not active in winter, dampers should be tightly closed during heating operation and when roof equipment is off.
- Demand limiting of central cooling equipment, and for VFD fans and pumps, to 85% capacity during peak demand periods.
- VAV—avoid supply air reset except in winter.
- VAV—avoid supply air reset that is based on return temperature.
- VAV—reduce supply duct static pressure.
- VAV—verify VAV box minimums are appropriately low and that the "heating minimum CFM" is no higher than the cooling minimum CFM.
- VAV—Classroom and lecture hall occupancy sensor to shut off HVAC systems when unoccupied for over 15 minutes. Perimeter rooms VAV HVAC goes to minimum air flow, interior rooms fully closed off.

Maintenance
- Calibrate controls every two years.
- Annual maintenance on all heat transfer surfaces, including good quality filters, cleaning coils, cleaning tubes.
- Use cog belts instead of standard V-belts for large motors.
- Seal return air plenums against any leakage to outside. Verify plenum temperature is within 2 degrees of room temperature in summer and winter design days.

Low Cost/No Cost
- Programmable thermostat for DX HVAC equipment.
- Install solar screening over large skylights and sun rooms.
- Occupancy sensors for lighting in classrooms, lecture hall, and labs.
- Photo cell control of parking lot and exterior lighting.
- Global control for "computer monitors off" after 15 minutes of inactivity, instead of screen savers.

Retrofit—Or Upgrade At Normal Replacement
Heating and Cooling
- Higher efficiency heating and cooling equipment.

- Convert constant volume HVAC to VAV.
- VFDs replace inlet vanes on air handlers.
- Evaporative pre-cooling on large air cooled package rooftop units.
- Anti-stratification fans for high bay, heated areas to move warm air to the floor.
- Dedicated cooling system for 24x7 needs in small areas, to allow the main building HVAC system to shut off at night.
- Reduce outside air to proper quantities if excessive.
- Increase roof insulation thickness to current energy code level as part of roof replacement.

Lighting
- Higher efficiency lighting.
- Replace incandescent lights with more efficient technology.
- Reduce excessive light levels by de-lamping.
- Combine de-lamping and reflectors to maintain light levels.

May Only Be Viable During New Construction
- Light color exterior walls.
- Light color high emissivity roof for low rise buildings.
- Passive shading elements for large glass areas.
- High performance low-E glazing.
- Exhaust heat recovery for high ventilation areas such as labs.
- Replace flat filters with angled filters.
- Circuiting of lights to allow first 10 feet inboard from the perimeter to be turned off during bright outdoor hours.
- Daylight lighting design.

Dormitories: See "Apartments" in Part A.
Cafeteria: See "Food Service" in Part A.
If a boiler is used: See "Boilers" in Part A.
If a chiller is used: See "Chillers" in Part A.
If the building is over four stories in height: See "High Rise" in Part A.

EDUCATION—SCHOOLS K-12

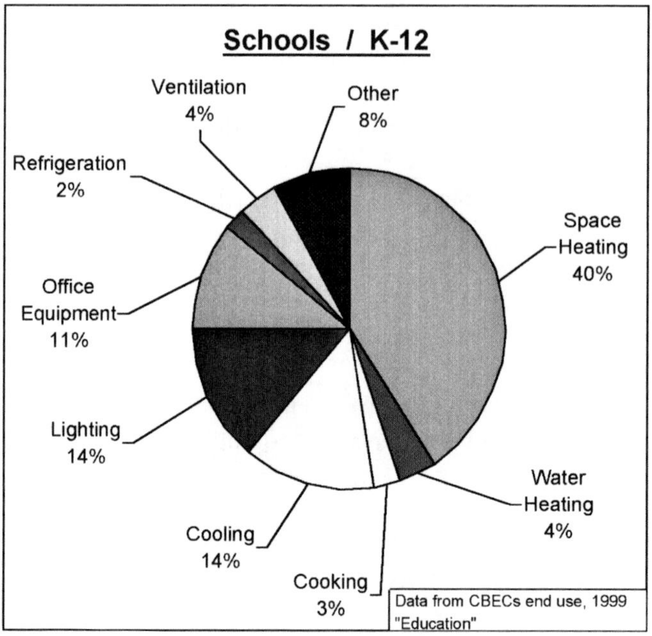

Schools / K-12

Ventilation 4%
Other 8%
Refrigeration 2%
Office Equipment 11%
Lighting 14%
Cooling 14%
Cooking 3%
Space Heating 40%
Water Heating 4%

Data from CBECs end use, 1999 "Education"

Figure 3A-4. Energy End Use Distribution for Education—Schools K-12

Primary Energy Use Sources
- Heating and cooling
- Lighting
- Office equipment

Controls
- Scheduled start-stop of HVAC equipment and lighting.
- Set space temperatures to 75 cooling/70 heating.
- Set space temperatures during unoccupied times up to 85/set back to 60 at night.
- Lock out cooling below 55 degrees and use only economizer.
- Lock out heating above 65 deg F.
- Extend the economizer cooling function, from 55 deg F to 60 deg F, if outdoor dew point levels are at or below indoor dew point levels— i.e. if not too humid outdoors.
- 5 degree deadband between heating and cooling operations.

- Eliminate simultaneous heating and cooling.
- Unit ventilators use ASHRAE Control Cycle 2.
- For all roof openings to hoods and equipment not active in winter, dampers should be tightly closed during heating operation and when roof equipment is off.
- Demand controlled ventilation in assembly areas and other large controllable variable occupancy areas.
- VAV—avoid supply air reset except in winter.
- VAV—avoid supply air reset that is based on return temperature.
- VAV—reduce supply duct static pressure.
- VAV—verify VAV box minimums are appropriately low and that the "heating minimum CFM" is no higher than the cooling minimum CFM.
- VAV—Classroom occupancy sensor to shut off HVAC when unoccupied for over 15 minutes. Perimeter rooms VAV HVAC goes to minimum air flow, interior rooms fully closed off.

Maintenance
- Calibrate controls every two years.
- Annual maintenance on all heat transfer surfaces, including good quality filters, cleaning coils, cleaning tubes.
- Seal return air plenums against any leakage to outside. Verify plenum temperature is within 2 degrees of room temperature in summer and winter design days.

Low Cost/No Cost
- Programmable Thermostat for DX HVAC equipment.
- Occupancy sensors for lighting in classrooms and labs.
- Photo cell control of exterior lighting.

Retrofit—Or Upgrade At Normal Replacement
Heating and Cooling
- Higher efficiency heating and cooling equipment.
- Reduce outside air to proper quantities if excessive.
- Increase roof insulation thickness to current energy code level as part of roof replacement.

Lighting
- Higher efficiency lighting.

- Replace incandescent lights with more efficient technology.
- Reduce excessive light levels by de-lamping.
- Combine de-lamping and reflectors to maintain light levels.

May Only Be Viable During New Construction
- Light color exterior walls.
- Light color high emissivity roof for low rise buildings.
- Passive shading elements for large glass areas.
- High performance low-E glazing.
- Exhaust heat recovery for high ventilation areas such as labs.
- Replace flat filters with angled filters.
- Circuiting of lights to allow first 10 feet inboard from the perimeter to be turned off during bright outdoor hours.
- Daylight lighting design.
- Ground source heat pumps.

Cafeteria: See "Food Service" in Part A.
If a boiler is used: See "Boilers" in Part A.
If a chiller is used: See "Chillers" in Part A.

FOOD SALES—GROCERY STORES

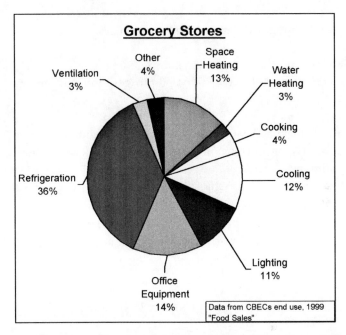

Figure 3A-5. Energy End Use Distribution for Food Sales/Grocery Stores

Primary Energy Use Sources
- Refrigeration
- Office equipment
- Heating and cooling
- Lighting

Controls
- Demand controlled ventilation for variable occupancy
- Set back temperatures if not open 24 hours.

Maintenance
- Annual maintenance on all heat transfer surfaces, including good quality filters, cleaning coils, cleaning tubes.
- Calibrate HVAC controls every two years.
- Calibrate refrigeration controls every year.

Low Cost/No Cost
- Air curtains at customer entrance and shipping docks.

Retrofit—Or Upgrade At Normal Replacement
Refrigeration
- Higher efficiency refrigeration equipment.
- Convert air-cooled to water-cooled refrigeration equipment.

Heating and Cooling
- Higher efficiency heating and cooling equipment
- Evaporative pre-cooling on large air cooled package rooftop units in dry climates.
- Increase roof insulation thickness to current energy code level as part of roof replacement.

Lighting
- Higher efficiency lighting.

Other
- Power factor correction for groups of large motors.

May Only Be Viable During New Construction
- Heat recovery from refrigeration equipment for space heat or dehumidification.
- Light color high emissivity roof for low rise buildings.
- Daylight lighting design.

HEALTH CARE—HOSPITAL

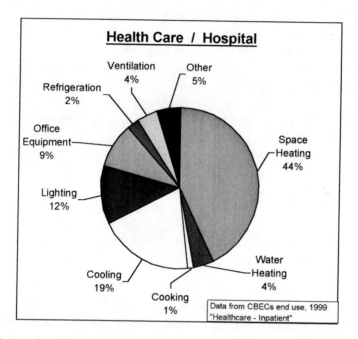

Figure 3A-6. Energy End Use Distribution for Health Care—Hospital

Primary Energy Use Sources
* Heating and cooling
* Lighting
* Water heating

Controls
* Space temperatures set to 75 cooling/70 heating in live-in areas.
* Set space temperatures during unoccupied times up to 85/set back to 60 at night, for staff areas not continuously occupied.
* 5 degree deadband between heating and cooling operations.
* Eliminate simultaneous heating and cooling.
* Use outdoor economizer instead of mechanical cooling below 55 deg F, if prescribed space pressurization can be maintained.
* Extend the economizer cooling function, from 55 deg F to 60 deg F, if outdoor dew point levels are at or below indoor dew point levels— i.e. if not too humid outdoors.

- Constant volume HVAC—supply air reset from demand to reduce reheat.
- VAV—avoid supply air reset except in winter.
- VAV—avoid supply air reset that is based on return temperature.
- VAV—reduce supply duct static pressure.
- VAV—verify VAV box minimums are appropriately low and that the "heating minimum CFM" is no higher than the cooling minimum CFM.
- VAV—Meeting rooms and day-time office areas—occupancy sensor to shut off HVAC when unoccupied for over 15 minutes. Perimeter rooms VAV HVAC goes to minimum air flow, interior rooms fully closed off.
- Avoid humidification if possible.
- Lower domestic hot water temperature to 120 deg F for hand washing and bathing.
- Turn off domestic hot water re-circulation pumps at night to unoccupied areas of the building.

Maintenance
- Calibrate controls every two years.
- Annual maintenance on all heat transfer surfaces, including good quality filters, cleaning coils, cleaning tubes.
- Use cog belts instead of standard V-belts for large motors.

Low Cost/No Cost
- Occupancy sensors for lighting in meeting rooms and day-time office areas.
- Insulate bare hot piping for domestic water.
- Low flow faucets.

Retrofit—Or Upgrade At Normal Replacement
Heating and Cooling
- Higher efficiency heating and cooling equipment.
- Heat recovery from exhaust air, where exhaust is continuous.
- VFDs replace inlet vanes on air handlers.

Water Heating
- Low flow shower heads.
- Condensing domestic water heater.
- Stack damper for gas-fired domestic water heater.

Lighting
- Higher efficiency lighting.
- Replace incandescent lights with more efficient technology.
- Reduce excessive light levels by de-lamping.
- Combine de-lamping and reflectors to maintain light levels.

Other
- Power factor correction for groups of large motors.

May Only Be Viable During New Construction
- Replace flat filters with angled filters.
- Water-side economizer to make chilled water with just cooling towers, if air economizers are not used.
- VAV—supply and return box tracking instead of constant volume.
- Separate the water heating systems for 120, 140, and 160 degree hot water.

If a boiler is used: See "Boilers" in Part A.
- Boiler stack gas heat recovery.

If a chiller is used: See "Chillers" in Part A.

If the building is over four stories in height: See "High Rise" in Part A.

HEALTH CARE—NON-HOSPITAL

Primary Energy Use Sources
- Heating and cooling
- Cooking (live-in facilities)
- Lighting

Controls
- Scheduled start-stop of HVAC equipment and lighting.
- Set space temperatures to 75 cooling/70 heating. Avoid over-cooling.
- Set space temperatures during unoccupied times up to 85/set back

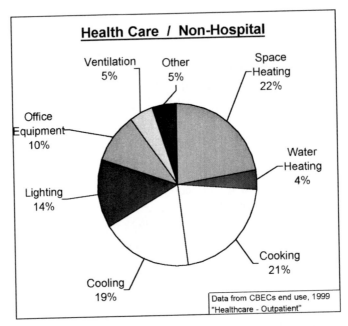

Figure 3A-7. Energy End Use Distribution for Health Care—Non-hospital

to 60 at night.

- 5 degree deadband between heating and cooling operations.
- Eliminate simultaneous heating and cooling.
- Lock out cooling below 55 degrees and use only economizer.
- Lock out heating above 65 deg F.
- Extend the economizer cooling function, from 55 deg F to 60 deg F, if outdoor dew point levels are at or below indoor dew point levels— i.e. if not too humid outdoors.
- Optimum start and stop of primary cooling and heating equipment.
- Constant volume HVAC—supply air reset from demand to reduce reheat.
- VAV—avoid supply air reset except in winter.
- VAV—avoid supply air reset that is based on return temperature.
- VAV—reduce supply duct static pressure.
- VAV—verify VAV box minimums are appropriately low and that the "heating minimum CFM" is no higher than the cooling minimum CFM.
- Avoid humidification if possible.
- Eliminate simultaneous humidification/dehumidification

- Demand limiting of central cooling equipment, and for VFD fans and pumps, to 85% capacity during peak demand periods.
- Demand controlled ventilation in office areas.

Maintenance
- Calibrate controls every two years.
- Annual maintenance on all heat transfer surfaces, including good quality filters, cleaning coils, cleaning tubes.
- Seal return air plenums against any leakage to outside. Verify plenum temperature is within 2 degrees of room temperature in summer and winter design days.

Low Cost/No Cost
- Programmable thermostat—use set up/set back feature at night unless serving sensitive equipment.
- Occupancy sensors for lighting in meeting rooms and restrooms.

Retrofit—Or Upgrade At Normal Replacement
Heating and Cooling
- Higher efficiency heating and cooling equipment.
- Dedicated small cooling system for 24x7 needs in small areas, to allow the main building HVAC system to shut off at night.
- Reduce outside air to proper quantities if excessive.
- Increase roof insulation thickness to current energy code level as part of roof replacement.

Lighting
- Higher efficiency lighting.
- Replace incandescent lights with more efficient technology.
- Reduce excessive light levels by de-lamping.
- Combine de-lamping and reflectors to maintain light levels.

May Only Be Viable During New Construction
- Light color high emissivity roof for low rise buildings.
- Replace flat filters with angled filters.

Kitchen: See "Food Service" in Part A.

LAUNDRIES—COMMERCIAL

Primary Energy Use Sources
- Water heating
- Heat for tumble drying
- Steam for pressing

Controls
- Turn off domestic hot water re-circulation pumps at night.

Maintenance
- Annual cleaning of heat transfer surfaces, including water heaters, heat recovery equipment.

Low Cost/No Cost
- Insulate bare hot piping.
- Reduce water temperature if possible.

Retrofit—Or Upgrade At Normal Replacement
Water Heating
- Heat recovery for waste water, to preheat cold water.
- Heat recovery for hot exhaust from dryers and roller irons, to preheat cold water.
- Condensing water heater.
- Stack damper for water heater.

Cooling
- Evaporative cooling.

May Only Be Viable During New Construction
- Washers that use less water.
- High speed extractors.

Boiler: See "Boilers" in Part A.

LIBRARIES/MUSEUMS

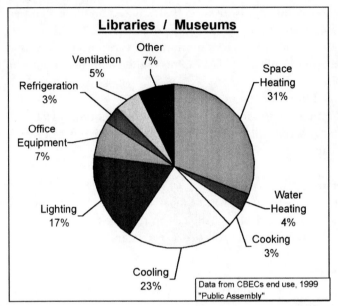

Libraries / Museums

Other 7%

Ventilation 5%

Refrigeration 3%

Office Equipment 7%

Lighting 17%

Cooling 23%

Space Heating 31%

Water Heating 4%

Cooking 3%

Data from CBECs end use, 1999 "Public Assembly"

Figure 3A-8. Energy End Use Distribution for Libraries/Museums

Primary Energy Use Sources
- Heating and cooling
- Lighting

Controls
- Scheduled start-stop of lighting even if HVAC must run continuously.
- Set space temperature controls to avoid over-cooling.
- 5 degree deadband between heating and cooling operations if possible.
- Eliminate simultaneous heating and cooling.
- If air economizer is not allowed for humidity stabilization, use water economizer or dry cooler below 55 degrees outside temperature.
- Lock out cooling below 50 degrees F.
- Lock out heating above 65 deg F unless dehumidification cycle is needed.
- Constant volume HVAC—supply air reset from demand to reduce reheat.

- VAV—avoid supply air reset except in winter.
- VAV—avoid supply air reset that is based on return temperature.
- VAV—reduce supply duct static pressure.
- VAV—verify VAV box minimums are appropriately low and that the "heating minimum CFM" is no higher than the cooling minimum CFM.
- Avoid humidification if possible.
- Eliminate simultaneous humidification/dehumidification.
- Demand controlled ventilation for unoccupied periods and variable occupancy.

Maintenance
- Annual maintenance on all heat transfer surfaces, including good quality filters, cleaning coils, cleaning tubes.
- Calibrate HVAC controls every two years.

Low Cost/No Cost
- Light sensors to harvest daylight near skylights and reading areas near glass.

Retrofit—Or Upgrade At Normal Replacement
Heating and Cooling
- Higher efficiency heating and cooling equipment.
- Install solar screening over large skylights and sun rooms.
- Install evaporative pre-cooling on air-cooled chillers in dry climates.

Lighting
- Higher efficiency lighting.
- Replace incandescent lights with more efficient technology.
- Reduce excessive light levels by de-lamping.

May Only Be Viable During New Construction
- Light color high emissivity roof for low rise buildings.

Boiler: See "Boilers" in Part A.
Chiller: See "Chillers" in Part A.

LODGING/HOTELS/MOTELS

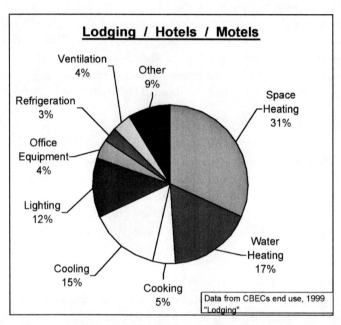

Figure 3A-9. Energy End Use Distribution for Lodging/Hotels/Motels

Primary Energy Use Sources
- Heating and cooling
- Water heating
- Lighting

Controls
- Scheduled start-stop of HVAC equipment and lighting in staff areas and other areas without continuous guest access.
- Set space temperatures to 75 cooling/70 heating.
- Set space temperatures during unoccupied times up to 85/set back to 60 at night, for staff areas not continuously occupied and amenity areas without continuous guest access.
- Lock out cooling below 55 degrees and use only economizer (common and amenity areas).
- Lock out heating above 65 deg F (common and amenity areas).
- Extend the economizer cooling function, from 55 deg F to 60 deg F, if

outdoor dew point levels are at or below indoor dew point levels—
i.e. if not too humid outdoors.
- Eliminate overlapping heating and cooling in common area HVAC
  equipment by adjusting control settings to have a "deadband."
- Lower domestic hot water temperature to 120 deg F for hand washing
  and bathing, and 140 deg F for dish washing, if possible.
- Turn off domestic hot water re-circulation pumps at night.
- Guest occupancy controls to shut off lights and set back HVAC.

Larger Facilities
- Demand controlled ventilation for variable occupancy areas such as
  meeting rooms, ballrooms, and common areas.
- Avoid over-cooling large meeting rooms and ballrooms. Limit these
  space temperatures to 70 degrees.
- Demand limiting of central cooling equipment, and for VFD fans
  and pumps, to 85% capacity during peak demand periods.

Maintenance
- Calibrate controls every two years.
- Annual maintenance on all heat transfer surfaces, including good
  quality filters, cleaning coils, cleaning tubes.
- Use cog belts instead of standard V-belts for large motors.
- Seal return air plenums against any leakage to outside. Verify plenum
  temperature is within 2 degrees of room temperature in summer and
  winter design days.
- Repair or replace defective zone control valves.

Low Cost/No Cost
- Insulate bare hot piping for domestic water.
- Install solar screening over large skylights and sun rooms.
- Occupancy sensors for lighting in amenity areas and meeting
  rooms.
- Close PTAC "vent" function if corridor ventilation is adequate.

Retrofit—Or Upgrade At Normal Replacement
Heating and Cooling
- Higher efficiency heating and cooling equipment.
- Higher efficiency guest room package terminal unit air conditioners
  (PTACs).

- In mild climates, heat pump PTACs instead of electric heat.
- Increase roof insulation thickness to current energy code level as part of roof replacement.

Water Heating
- Low flow shower heads.
- Stack damper for gas-fired domestic water heater.
- Condensing domestic water heater.

Lighting
- Higher efficiency lighting.
- Replace incandescent lights with more efficient technology.
- Reduce excessive light levels by de-lamping.
- Combine de-lamping and reflectors to maintain light levels.

Larger Facilities
- Install solar screening over large skylights and sun rooms.
- Convert constant volume HVAC to VAV.
- VFDs replace inlet vanes on air handlers.
- Power factor correction for groups of large motors.
- Dedicated cooling system for 24x7 needs in small areas, to allow the main building HVAC system to shut off at night.
- Reduce outside air to proper quantities if excessive.
- Low water volume/high speed spin wash machines (Energy Star).

May Only Be Viable During New Construction
- Envelope construction in guest areas to match local residential energy code for insulation.
- Overhangs or other exterior shading at windows.
- Light color high emissivity roof for low rise buildings.
- Toilet exhaust on demand (light switch) instead of continuous.
- Low volume bathtubs.
- Separate the water heating systems for 120, 140, and 160 degree hot water.
- Gas-fired PTACs instead of electric resistance.
- Ground source heat pumps.

Kitchen: See "Food Service" in Part A.
Pool: See "Pools" in Part A.

If a boiler is used: See "Boilers" in Part A.
If a chiller is used: See "Chillers" in Part A.
If the building is over four stories in height: See "High Rise" in Part A.

OFFICE BUILDINGS

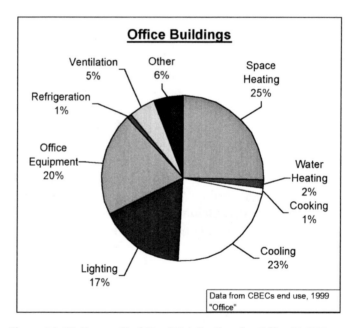

Figure 3A-10. Energy End Use Distribution for Office Buildings

Primary Energy Use Sources
• Heating and cooling
• Lighting
• Office equipment

Controls
• Scheduled start-stop of HVAC equipment and lighting.
• Set space temperatures to 75 cooling/70 heating.
• Set space temperatures during unoccupied times up to 85/set back

to 60 at night.
- 5 degree deadband between heating and cooling operations.
- Eliminate simultaneous heating and cooling.
- Lock out cooling below 55 degrees and use only economizer.
- Lock out heating above 65 deg F.
- Extend the economizer cooling function, from 55 deg F to 60 deg F, if outdoor dew point levels are at or below indoor dew point levels— i.e. if not too humid outdoors.
- VAV—avoid supply air reset except in winter.
- VAV—avoid supply air reset that is based on return temperature.
- VAV—reduce supply duct static pressure.
- VAV—Verify VAV box minimums are appropriately low and that the "heating minimum CFM" is no higher than the cooling minimum CFM.

Larger Facilities
- Optimum start and stop of primary cooling and heating equipment.
- Stage electric heating to prevent setting seasonal demand and invoking ratchet charges.
- VAV—Conference room occupancy sensor to shut off HVAC when unoccupied for over 15 minutes. Perimeter rooms VAV HVAC goes to minimum air flow, interior rooms fully closed off.
- Demand controlled ventilation for variable occupancy areas such as conference rooms and open plan office areas.
- Demand limiting of central cooling equipment, and for VFD fans and pumps, to 85% capacity during peak demand periods.

Maintenance
- Calibrate controls every two years.
- Annual maintenance on all heat transfer surfaces, including good quality filters, cleaning coils, cleaning tubes.
- Use cog belts instead of standard V-belts for large motors.
- Seal return air plenums against any leakage to outside. Verify plenum temperature is within 2 degrees of room temperature in summer and winter design days.

Low Cost/No Cost
- Programmable thermostat for package A/C units—set up to 85 deg/set back to 60 deg at night.

- Install solar screening over large skylights and sun rooms.
- Occupancy sensors for lighting in conference rooms and meeting rooms.
- Global control for "computer monitors off" after 15 minutes of inactivity, instead of screen savers.

Retrofit—Or Upgrade At Normal Replacement
Heating and Cooling
- Higher efficiency heating and cooling equipment.
- Convert constant volume HVAC to VAV.
- VFDs replace inlet vanes on air handlers.
- Reduce outside air to proper quantities if excessive.
- Increase roof insulation thickness to current energy code level as part of roof replacement.

Lighting
- Higher efficiency lighting.
- Replace incandescent lights with more efficient technology.
- Reduce excessive light levels by de-lamping.
- Combine de-lamping and reflectors to maintain light levels.

Larger Facilities
- Evaporative pre-cooling on large air cooled package rooftop units in dry climates.
- Power factor correction for groups of large motors.
- Dedicated cooling system for 24x7 needs in small areas, to allow the main building HVAC system to shut off at night.
- Window film or light colored interior shades for buildings with large amounts of glass.

May Only Be Viable During New Construction
- High performance low-E glazing
- Overhangs or exterior shading for glazing.
- Light color exterior walls.
- Light color high emissivity roof for low rise buildings.
- Circuiting of lights to allow first 10 feet inboard from the perimeter to be turned off during bright outdoor hours.

Data center: See "Data Centers" in Part A.
If a boiler is used: See "Boilers" in Part A.

If a chiller is used: see "Chillers" in Part A.
If the building is over four stories in height: See "High Rise" in Part A.

FOOD SERVICE/RESTAURANTS

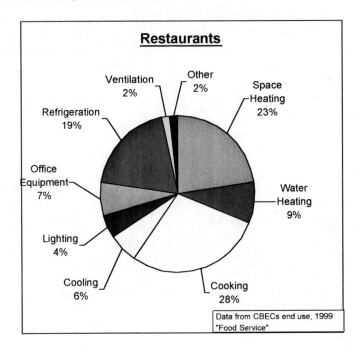

Figure 3A-11. Energy End Use Distribution for Food Service/Restaurants

Primary Energy Use Sources
- Cooking
- Heating and cooling
- Refrigeration
- Water heating

Kitchen and Equipment
- Run hoods only when needed.

- Control multiple hoods independently and provide variable demand-based make-up air.
- Turn down cooking equipment to standby temperature when not in use.
- Provide controls so the food warming lamps only operate when there is food there to be warmed, instead of always on.
- Limit kitchen cooling to 75 degrees.
- Limit freezer temperature to zero deg F—put ice cream in a separate small chest freezer.
- Limit cooler temperature to 37 deg F.
- Limit hood make-up air tempering to 50 degrees heating.
- Disable the heating for the HVAC units that serve the kitchen area, or lower the setting to only come on below 60 degrees. This is a "cooling only" application and should not require heating.
- Install interlock for the dishwasher hood exhauster so that it only runs when the dishwasher runs.
- Gaskets and door sweeps on walk-in cooler and freezer doors.
- Hanging plastic strips at openings into walk-in coolers and freezers.
- Locate condensing units to reject heat outside, for coolers, freezers, ice makers.
- Use chemical sanitizer rinse instead of 180 deg F booster heater.
- Variable flow heat hoods (Class 1)
- UL listed hoods for reduced air flows.
- Direct fired hood make-up heating.
- Separate hot and cold equipment used for food storage and preparation.
- Air balance to assure proper hood capture of hot cooking fumes.
- Walk-in freezer and cooler combined so entrance to freezer is from the walk-in cooler.
- On-demand cooking equipment such that energy use closely tracks production, and will allow proportionally reduced energy cost during slow times.

Controls
- Scheduled start-stop of HVAC equipment and lighting.
- Set space temperatures to 75 cooling/70 heating.
- Set space temperatures during unoccupied times up to 85/set back to 60 at night.
- Lock out cooling below 55 degrees and use only economizer.

- Lock out heating above 65 deg F.
- Lower domestic hot water temperature to 120 deg F for hand-washing, and 140 deg F for dish washing, if possible.
- Turn off domestic hot water re-circulation pumps at night.
- (Dining areas) demand controlled ventilation unless tied to kitchen exhaust/make-up system.

Maintenance
- Calibrate controls every two years.
- Each three months or less: maintenance on all heat transfer surfaces, including good quality filters, cleaning coils, cleaning tubes. Include refrigeration equipment, coolers, and freezers.

Low Cost/No Cost
- Programmable thermostat to prevent HVAC equipment from running continuously at night.
- Install solar screening over large skylights and sun rooms.
- Insulate bare hot piping for domestic water.

Retrofit—Or Upgrade At Normal Replacement
Heating and Cooling
- Add A/C economizer for kitchen area, if none exists.
- Higher efficiency cooling equipment.
- Down-size HVAC cooling equipment if over-sized.

Water Heating
- Stack damper for gas-fired domestic water heater.
- Condensing domestic water heater.

RETAIL/SALES

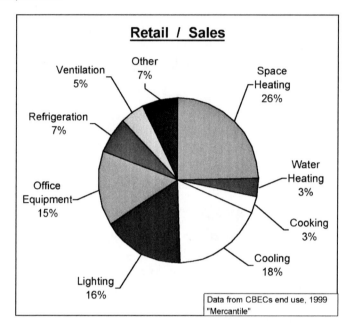

Figure 3A-12. Energy End Use Distributionfor Retail/Sales

Primary Energy Use Sources
- Heating and cooling
- Lighting
- Office equipment

Controls
- Scheduled start-stop of HVAC equipment and lighting.
- Set space temperatures to 75 cooling/70 heating.
- Set space temperatures during unoccupied times up to 85/set back to 60 at night.
- Lock out cooling below 55 degrees and use only economizer.
- Lock out heating above 65 deg F.
- Extend the economizer cooling function, from 55 deg F to 605 deg F, if outdoor dew point levels are at or below indoor dew point levels—i.e. if not too humid outdoors.
- Maintain a 5 deg F deadband between heating and cooling for all HVAC equipment.

- Turn off display lighting except during customer times.

Larger Facilities
- Optimum start and stop of primary cooling and heating equipment.
- Demand controlled ventilation for variable occupancy times.
- Photo cell control of parking lot lighting.
- For all roof openings to hoods and equipment not active in winter, dampers should be tightly closed during heating operation and when roof equipment is off.
- If demand metered: demand limiting of central cooling equipment, and for VFD fans and pumps, to 85% capacity during peak demand periods.

Maintenance
- Calibrate controls every two years.
- Annual maintenance on all heat transfer surfaces, including good quality filters, cleaning coils, cleaning tubes.

Larger Facilities
- Use cog belts instead of standard V-belts for large motors.
- Seal return air plenums against any leakage to outside. Verify plenum temperature is within 2 degrees of room temperature in summer and winter design days.

Low Cost/No Cost
- Programmable thermostat for package A/C units—set up to 85 deg/ set back to 60 deg at night.
- Occupancy sensors for lighting in break rooms.
- Install solar screening over large skylights.

Retrofit—Or Upgrade At Normal Replacement
Heating and Cooling
- Higher efficiency heating and cooling equipment.
- HVAC air economizer, lock out mechanical cooling below 55 deg F.
- Increase roof insulation thickness to current energy code level as part of roof replacement.

Lighting
- Higher efficiency lighting.

- Replace incandescent lighting with more efficient technology.
- Reduce excessive light levels by de-lamping.
- Combine de-lamping and reflectors to maintain light levels.

Larger Facilities
- High bay lighting retrofit, HID to fluorescent.
- Evaporative pre-cooling on large air cooled package rooftop units.
- Convert constant volume HVAC to VAV.
- VFDs replace inlet vanes on air handlers.
- Power factor correction for groups of large motors.
- Reduce outside air to proper quantities if excessive.

May Only Be Viable During New Construction
- Overhang over storefront window.
- HVAC variable volume air systems.
- Light color high emissivity roof for low rise buildings.
- Daylight lighting design.

If a boiler is used: see "Boilers" in Part A.
If a chiller is used: see "Chillers" in Part A.
If the building is over four stories in height: see "High Rise" in Part A.

WAREHOUSES

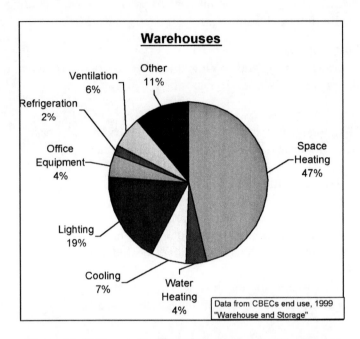

Figure 3A-13. Energy End Use Distribution for Warehouses

Primary Energy Use Sources
- Heating
- Lighting

Controls
- Occupancy sensors to turn off lights in unmanned areas and during unoccupied times.
- Set space temperatures to 80 cooling/60 heating unless product storage requires closer temperature control.
- Set space temperatures during unoccupied times up to 90/set back to 50 at night unless product storage requires closer temperature control.
- For all roof openings to hoods and equipment not active in winter, dampers should be tightly closed during heating operation and when roof equipment is off.

Maintenance
- Annual maintenance on all heat transfer surfaces, including good quality filters, cleaning coils.

Low Cost/No Cost
- Interlock heating and cooling at loading docks to stop when roll-up doors are opened.

Refrigerated Warehouse
- Hanging plastic strips at openings into refrigerated areas.
- Air curtains at openings into refrigerated areas.

Retrofit—Or Upgrade At Normal Replacement:
Heating and Cooling
- Higher efficiency heating equipment.
- Gas-fired radiant heating in lieu of space heating.
- Anti-stratification fans for high bay, heated areas to move warm air to the floor.
- Increase roof insulation thickness to current energy code level as part of roof replacement, for buildings that are heated or cooled.

Lighting
- High bay lighting retrofit HID to fluorescent.
- Higher efficiency lighting.
- Motion sensors for overhead lights, except re-strike time makes this impractical for HID lighting.
- Reduce excessive light levels by de-lamping.
- Combine de-lamping and reflectors to maintain light levels.

Refrigerated Warehouse
- Higher efficiency lighting to reduce refrigeration load.
- Convert air-cooled to water-cooled refrigeration.
- Increase roof insulation thickness over refrigerated areas.

May Only Be Viable During New Construction
- Improved insulation for buildings intended to be heated or cooled

Refrigerated Warehouse
- Improve envelope insulation

- Light color exterior walls for refrigerated areas.
- Light color high emissivity roof for refrigerated areas.
- Daylight lighting design.

BOILERS

- Controls to lock out boiler above 60 deg F.
- Reset boiler temperature from outside air.
- Annual verification that hydronic heating and cooling automatic control valves close tightly and prevent any internal leak-by.
- Modular boilers to reduce standby losses, upon replacement.
- Stack dampers for single boilers that cycle frequently during part load.
- Annual efficiency checking. Take corrective action if efficiency is found to be less than 95% of new equipment values.
- Automatic boiler isolation valves if piping allows hot water through an "off" standby boiler.
- Insulate bare heating water piping.
- Maintain proper water treatment and ensure that normal make-up does not cause dilution and scaling.
- Convert constant flow heating water to variable flow for larger systems with high annual run hours.
- Separate domestic water heating equipment, instead of using heating boiler during summer for this purpose.
- Insulate boiler surface areas and access panels that have no casing insulation.
- Reduce excess air for burner to 30% or less.

CHILLERS

- Higher efficiency chiller.
- Vary auxiliary pump flows in proportion to cooling load instead of constant flow.

- Lower condenser water temperature for water-cooled chiller.
- Higher efficiency/capacity cooling tower upon replacement. Suggested criteria for replacement cooling tower is a maximum design approach of 7 degrees to design wet bulb conditions and a fan power budget of no more than 0.05 kW/ton.
- VFDs for cooling tower fans.
- Annual condenser tube cleaning for water-cooled chiller.
- Controls to lock out the chiller below 50 deg F.
- Evaporative pre-cooling for air-cooled chillers.
- Raise chilled water temperature if possible, but use caution when dehumidification is needed.
- Convert constant flow chilled water to variable flow for larger systems with high annual run hours.
- Insulate bare chilled water piping.

HIGH RISE

Envelope
- Entry door vestibule or revolving door.
- Seal around punch windows.
- Seal vertical shafts.
- Seal air plenums.
- Motorized damper to close elevator shaft except in fire mode, if allowed by local building regulations.

POOLS

Primary Energy Use Sources
- Pool water heating
- Pool air heating and dehumidifying
- Shower water heating

Controls
- Maintain air temperature within 2 degrees above water temperature.
- Maintain air humidity above between 50-60%, do not over dry the air through excessive ventilation or other means.
- Reduce pool water temperature overall.
- Relax pool water temperature, air temperature, and humidity requirements during unoccupied times. Allow water to cool a few degrees at night if possible.
- Reduce air flow rates and air exchange rates in unoccupied periods.
- For all roof openings to hoods and equipment not active in winter, dampers should be tightly closed during heating operation and when roof equipment is off.

Maintenance
- Calibrate controls every two years.
- Annual efficiency checks for gas fired heaters.
- Annual maintenance on all heat transfer surfaces, including good quality filters, cleaning coils, cleaning tubes.
- Look for condensation on interior walls during winter that indicate poor insulation.

Low Cost/No Cost
- Drain outdoor pools in winter instead of heating.
- Minimize sand filter backwash, usually once per week is sufficient.

Retrofit—Or Upgrade at Normal Replacement
Pool Water Heating
- Condensing water heater.
- Filtration instead sand filters, to eliminate backwash and make-up water heating.
- Pool covers.

Pool Air Heating and Dehumidifying
- Reduce outside air to proper quantities if excessive.

Shower Water Heating
- Low flow shower heads.

May Only Be Viable During New Construction
- Replace mechanical cooling system with ventilation system in climates with acceptable summer humidity levels.
- For indoor pools, add a heating system for the surrounding air if there isn't one, to eliminate the pool from heating it.

ICE RINKS

Primary Energy Use Sources
- Refrigeration
- Space Heating
- Lighting

Controls
- Stage brine pumps proportionally with load.
- Control ice sheet temperature no lower than necessary.
- Control brine temperature no lower than necessary.
- Control condenser water temperature as low as possible to reduce head pressure.
- Turn off heaters in ice areas in unoccupied times.
- Closely monitor both temperature and humidity of outside air and only use outside air beyond minimums when beneficial.

Maintenance
- Calibrate controls every two years.
- Annual efficiency checks for gas fired heaters.
- Annual maintenance on all heat transfer surfaces, including good quality filters, cleaning coils, cleaning tubes.

Low Cost/No Cost
- Schedule ice-building times to use off-peak utility rates.
- Hanging barriers to keep warmed air for spectators from heating the ceiling over the ice slab.
- Window shades for glazing at ice slabs.
- Re-direct warm air heating supply to spectators but not at or over the ice sheet.

Retrofit—Or Upgrade at Normal Replacement
Refrigeration
• Replace conventional expansion valve with electronic expansion valve to allow colder condenser water, lower head pressures.
• Higher efficiency refrigeration equipment.
• Higher efficiency cooling tower with ample capacity (smaller fan motor)
• Variable speed brine pumping.
• Heat recovery from refrigeration system for space heating or for ice melting.
• Light colored roof over ice sheet.
• Increased wall and roof insulation around ice sheet.
• Increased insulation under ice sheet.
• Low-emissivity paint in interior roof over ice sheet.
• Heavy tint or frittered glazing near ice sheet.
• Switch to ethylene glycol if propylene glycol is being used.
• High efficiency lighting over ice sheet.
• Barriers to keep heated air over spectators from heating the ice sheet.

Space Heating
• Higher efficiency heating equipment
• Radiant heat for spectators instead of air heating.

Lighting
• Higher efficiency lighting has double effect by reduced load on the ice sheet.

PART B

Energy Conservation Measures
ECM Descriptions

ECM DESCRIPTIONS—BUILDING ENVELOPE

Envelope Leaks—Infiltration
Basis of savings: Heating and cooling unwanted outside air
- Infiltration from construction cracks usually manifests itself as a comfort complaint, but always increases energy use. In extreme cases it can result in frozen piping. A building pressure test is the perfect solution, but usually not practical. One easy way to check for infiltration leaks is to check return air plenum temperatures during very cold weather. The return plenum is a negatively pressurized area. If there is leakage, it can be found with a hand held infrared thermometer while scanning the perimeter above the ceiling tiles.
- Usually hard to quantify.
- In extreme cases, the leakage around old operable windows can represent more of an energy improvement opportunity than replacing the single pane windows with double pane windows.
- Return air ceiling and shaft plenums are intended to be as air tight as any other duct but are usually far from that. This is especially problematic at building perimeters when the lack of proper construction sealing couples the return air plenum to the building envelope. Since the return air plenum is slightly negative by design, this almost assures infiltration through the envelope. User complaints that point to this are comfort issues at perimeter and cold temperatures above the ceilings at the perimeter. In extreme cases, water pipes can freeze because of this. In humid climates, severe mold damage and sick buildings can be traced to this.

Exterior Color
Basis of savings: Reflecting, instead of absorbing heat.
- If the solar heat gain of the wall or roof can be identified, the difference in reflectivity represents the savings potential.

- For roofs, using a 'pure white' reflectance factor for long-term energy savings is not recommended, since the color will naturally darken with rain and dirt.

Insulation
Basis of savings: Reduced thermal transmission
- A rule of thumb is that 'The first inch of insulation captures 80% of the savings.'
- The biggest opportunities for retrofit are hot or cold surfaces with no insulation at all.
- Increasing insulation thickness during new construction is often cost effective since labor is similar and the material is not expensive.
- Adding insulation to a roof during a roof replacement that will occur anyway is usually cost effective.

Window Upgrade
Basis of savings: Reduced thermal transmission
- Glass is not a very good insulator, but a trapped air space does offer some insulation.
- In all but the most moderate of climates, single pane glazing is a poor choice from an energy perspective. In general, where heating or cooling is needed, single pane units should be avoided and are candidates for replacement if energy use reduction is desired.
- In addition to the glazing, some old window frames are a thermal short circuit. Evidence of this can often be found as water stains on the metal frames from winter condensation. New glazing should always have thermal breaks.
- For replacement of operable windows, poor seals and associated infiltration can pose as much of an energy loss as the windows themselves. New windows should have tight fitting seals.
- Windows are expensive and paybacks are commonly 15-20 years if only energy savings are used to justify the expense. However, window upgrades (the incremental upgrade cost) are easier to justify when windows are being installed anyway.

Window Shading
Basis of savings: Reduced solar load
- Exterior shades are best so that the heat never gets inside.
- Trees, awnings, overhangs, screen covers over skylights.

- Interior shades, if used, should be light colored and highly reflective.
- If existing glazing is clear, coatings or shades can reduce cooling load substantially and can reduce A/C equipment size requirements.

Light Harvesting
Basis of savings: Reduced interior lighting
- Portions of the solar load are deliberately allowed into the building to provide day lighting without the use of artificial light. Often this is done near the top of the room enclosure along the wall and reflected off the ceiling. Skylights and clerestories are used to harvest light from the roof to the room below.
- In all cases, the solar gain and added air conditioning load subtract from the lighting savings, although there is usually a net gain.
- For skylights and clerestories, transparent element acts like a hole in the insulation and increases heat loss at that point substantially, especially at night. The envelope heat loss adds heating load and subtracts from the lighting savings. Covers can mitigate this, but are cumbersome and expensive.

ECM DESCRIPTIONS—LIGHTING

Lighting Retrofit: T-12 Magnetic to T-8 Electronic Ballast
Basis of savings: Higher efficacy and reduced ballast loss (lumens per watt)
- Savings of 30% compared to T-12/magnetic ballast are common.
- This is a very common retrofit, since T-12 lights have been the staple fluorescent light installed for a number of years.
- Changing the ballast from "coil and core" (magnetic ballast) to electronic ballast represents the largest part of the savings.
- T-8 bulbs have better quality of light also, in terms of color rendition.
- Proposals to change the ballast but leave the T-12 lights are not recommended, due to compatibility issues. The bulbs need to be changed to T-8 along with the ballast.

Lighting Retrofit: Super T-8 with Low Factor Electronic Ballast

Basis of savings: Higher efficacy

- Special phosphors of these tubes result in greater light output. Without simultaneously reducing the ballast, this system results in the same energy use as a T-8 system with more light.
- But, by carefully pairing these high output bulbs with low "ballast factor" (BF), savings over conventional T-8 lamps with equal light are possible. For example, if the Super T-8 lamps put out 15% more light, and are paired with a ballast selected at BF=0.85 instead of 1.0, there will be approximately equivalent light and 15% less energy consumed, compared to standard T-8 systems.
- Super T-8 bulbs cost more than regular T-8s.

Lighting Retrofit: High Bay HID to High Bay Fluorescent

Basis of savings: Higher efficacy and reduced ballast loss (lumens per watt)

- HID = high intensity discharge
- Savings of 30-50% are common.
- Use caution for fluorescent lights in gymnasium areas since bulbs can shatter.
- High bay fluorescents require reflectors and may be problematic in dirty areas unless the reflectors can be wiped down regularly.
- This is a fixture replacement, usually 1-for-1 to re-use the electrical source.
- Can be T-5HO or, in some cases, T-8.

Lighting Retrofit: Incandescent to Compact Fluorescent (CFL)

Basis of savings: Higher efficacy

- Savings are typically 1/3 to 1/4 the energy for equivalent light, compared to incandescent.
- Installation of self-ballasted CFLs in ceiling 'down lights' is not recommended by manufacturers, since the heat build up will cause premature failure of most standard CFL ballasts. This application could provide reasonable service life if the recessed cans are vented.
- Generally not good for dimming, although advancements continue toward this goal.

Reflectors

Basis of savings: De-lamping with equivalent light

- All light fixtures result in some of the light being "trapped" in the fixture. Optical reflectors are available as retrofits. Made from highly reflective materials, these push more of the light out of the fixture. In many cases one of the tubes of an existing fixture can be eliminated by virtue of the reflectors. For example, a 3-lamp fixture, retrofitted with reflectors, may provide adequate light with the center tube removed.

- Be certain that the reflector unit is UL listed for a retrofit application.

De-lamping

Basis of savings: Reduced number of lamps reduces wattage in over-lit areas.

- If it is determined that light levels can be reduced, de-lamping is the simplest of all lighting measures.
- Magnetic ballast may or may not save the amount of energy implied by removing a tube. For example removing 1 of 3 tubes may not reduce energy use by 1/3.
- Most electronic ballasts are capable of operating with one less tube with no harm. If removing more than one tube, consult the ballast manufacturer to be sure. Power reduction with electronic ballast is roughly equivalent to the tubes removed.

ECM DESCRIPTIONS—HVAC

Seal Air Duct Leaks

Basis of savings: Reduced fan energy for a given air flow to the space. For leaks outside the conditioned space, savings are also from reduced unwanted outside air that is heated and cooled.

- Never overlook the obvious.
- In cooling season on a warm day the plenum temperature should be slightly warmer than the space below the ceiling. If it is found to be cooler in the return plenum than in the space on a warm day, then there is almost surely a duct leak.
- Duct sealing is especially important in unconditioned spaces (attics, basements) since air leaked at these points is truly lost. Duct leaks within the insulated envelope are not as critical since the heating or cooling energy is still there providing some benefit during the season.

Correct Control Valves Leaking By Internally
Basis of savings: Eliminating unwanted heating and cooling. In most cases, leaking control valves result in overlapping heating and cooling which doubles the energy waste.

- Testing consists of creating a "full close" command to the device and seeing if the downstream piping or coil returns to ambient or has measurable heating. Sometimes the first row of the coil will be found to be warmer or cooler than the rest of the coil, indicating leaks.
- Electronically actuated valves are especially prone to this due to adjustable travel stops that do not always have good residual close off seating pressure.
- For small piping, 1 inch and less, quarter-turn ball valves can be cost effective and have improved seating quality compared to conventional metal seated globe valves.

Insulate Piping and Valves
Basis of savings: Reduced thermal losses.

- Applies to both hot and chilled water systems, although hot systems represent the higher delta-T and thus the higher heat loss.
- For heating piping, safety is an added justification since many of these represent scald hazards.
- For chilled piping, corrosion protection is an added justification since the condensation that accompanies the cold surface temperature accelerates rust damage.

Lower Chilled Water Condensing Temperature
Basis of savings: Reduced refrigeration cycle "lift"

- 1-1.5% reduction in kW per degree lowered.

Raise Chilled Water Evaporating Temperature
Basis of savings: Reduced refrigeration cycle "lift"

- 1-1.5% reduction in kW per degree raised.

Air-side Economizer
Basis of savings: Avoided refrigeration compressor run time.

- Savings benefit varies by location (available hours of cool air) and by internal load characteristics (hours when cooling is needed while it is coincidentally cool outside).
- The outside air damper and relief dampers should be tightly closed whenever the equipment is off, and relief damper should be closed

tightly in winter when minimum outside air is used, to avoid infiltration in cold weather through these large dampers.

• For economizer mixing boxes that include a relief damper, automatically control it to be tightly closed whenever the outside air damper is at its minimum position, and only begin opening once the outside air damper position moves beyond minimum, lagging behind the outside air damper travel.

• There should be a setting for the economizer operation (usually 55-60 degrees F), below which outside air is deemed sufficient for any needed cooling. Below this point, there should be a positive "cooling lockout" function that prevents compressor operation or forces the chilled water valve fully closed.

Water-side Economizer
Basis of savings: Cooling loads and hours concurrent with low ambient wet bulb temperature can be met with the evaporative effect of the cooling tower directly, without running the chiller.

• The best use for these is if there are steady cooling loads in winter that cannot be served with air-side economizers, often due to the seasonal outside air humidity swings.

• These are inherently not as efficient as an air-side economizer since their use depends on pumps and cooling towers as well as the air handler. However, the chiller does get to shut off so there are substantial savings.

The trouble with most flat plate heat exchanger applications is that

• They are expensive.

• Their capacity is highest when indoor cooling load are usually lowest.

• They are arranged as all-or-nothing, so they are switched off even when they could provide most of the load.

• They share the chilled water system with "non-critical" chilled water loads. Without proper controls (namely, below 55 degrees, no chilled water to these units), the system reaches the cut off point sooner than it would need to.

• Note that there are some piping and pumping arrangements that will allow the flat plate system to run concurrently with the chiller, normally to pre-cool the chilled water return. This is a complex design solution but may be effective if the load and wet bulb temperature

CONVENTIONAL AND EXTENDED FLAT PLATE OPERATION					
	All Possible Flat Flat Plate Hours	approx. pct full load capacity	6a-6p annual hours	24-7 annual hours	Outside Air Conditions
EXTENDED	2.5 deg pre-cool	25%	180	480	OA 40-42.5 wb
EXTENDED	5 deg pre-cool	50%	240	630	OA 37.5-40 wb
EXTENDED	7.5 deg pre-cool	75%	60	150	OA 35-37.5 wb
CONVENTIONAL	chillers OFF	100%	1260	2820	OA <35 wb and <55 db
	increased flat plate hours		**38%**	**45%**	

Based on Wet Bulb Temperature Profile, Colorado Springs.

Figure 3B-1. Water-side Economizer Flat Plate Hours for One City

profiles show potential by stretching out the flat plate cutoff point. Sharing a single cooling tower, this requires some form of compensation for the chiller, since it cannot run on excessively cold condenser water. The accommodation may be throttling the condenser water flow to maintain head pressure on the chiller, or with recirculating/blending pump at the chiller condenser water inlet to temper the cold water. Where additional pumps are used to accommodate dual operation, the benefit of the partial flat plate operation must be compared with any additional pumping energy expense.

Angled Filters Instead of Flat Filters
Basis of savings: Reduced average pressure drop reduces air horsepower in VAV systems and in CAV systems if re-balanced.
- Can be combined with early change out for increased benefit.

Bag Filters instead of Cartridge Filters
Basis of savings: Reduced average pressure drop reduces air horsepower in VAV systems and in CAV systems if re-balanced.
- Can be combined with early change out for increased benefit.
- Filter manufacturers caution on "dirt release" from bag filters when stopped and started regularly, therefore this measure is best suited for fans with extended run times.

Multi-zone Conversion to VAV
Basis of savings: Reduced overlapping heating and cooling and reduced energy transport work since air flow is proportional to load instead of constant volume.
- Hot deck is blanked off and zone mixing dampers sealed in the "cold

deck" position. VAV boxes added for each zone, usually near the air handler and re-use the zone ductwork.
- A sketch of this modification is shown opposite.

Multi-zone: VAV Conversion Using Existing Zone Dampers
Basis of Savings: Reduced heating-cooling overlap and reduced fan energy.
- Main fan is controlled on static pressure in the hot/cold deck plenum box.
- Existing zone dampers act as pressure dependent VAV dampers. Uses existing ducts for economy.
- Linkage between hot and cold dampers must be split and actuators provided for independent hot and cold damper control.
- Unless the zone mixing dampers are in good condition and tight sealing, it may be more cost effective to convert to VAV.
- Care must be used to provide minimums and ventilation air.

Multi-zone Conversion to Texas Multi-zone
Basis of savings: Reduced overlapping heating and cooling penalty due to the introduction of the "bypass" air path.
- This conversion makes the hot deck of a multi-zone unit into a by-pass deck.
- Reheat coils are added for each zone downstream of the mixing dampers.
- The air stream delivered to the zones can be either cooled, warmed, or bypass (re-circulated), compared to either warmed or cooled.
- System remains constant volume.

Dual Duct Conversion to Separate Hot Deck and Cold Deck Fans
Basis of savings: Reduced heating burden in the hot deck.
- Independent control of hot duct and cold duct in the zone mixing boxes reduces the inherent overlap in these systems.
- Independent hot and cold duct systems via separate fans allow seasonal optimization for hot and cold decks. For example, mixing outside air to reduce cooling cost in winter adds to heating cost since both ducts share the same mixed air stream.
- Introduce the ventilation air into only one of the ducts—normally the cold duct.
- The hot deck simply re-circulates and does not see the cold ventilation air as a load.
- Additional savings are possible by converting from constant volume

to variable volume in both ducts.
- A sketch of this modification is shown below.

Spot Cooling
Basis of savings: Heat only the worker or process, not the whole factory.
- Concepts of "air changes" and "cfm per SF" do not apply.
- Duct size, cfm, and heat/cool energy delivered is much smaller.
- Can be very beneficial for un-insulated buildings, high bay buildings, and factories with a lot of stationary heat producing equipment

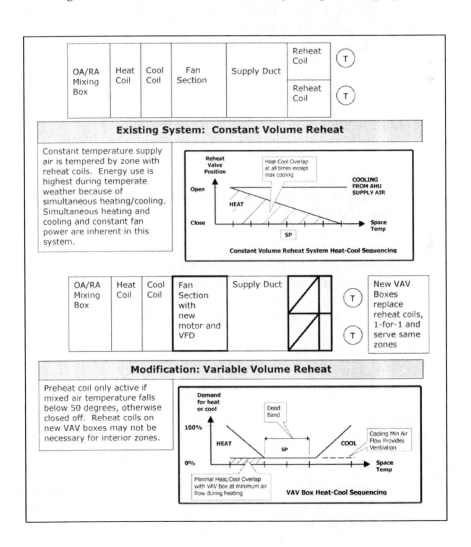

(ovens, kilns, laundry, etc).

- Special design considerations for this system to be effective.

Evaporative Pre-cooling for Air-cooled Condensers

Basis of savings: Reduction of dry bulb temperature from the evaporative cooling process lowers the condensing temperature and reduces refrigeration cycle lift. The air-cooled equipment "thinks" it is cooler outside and behaves accordingly.

- Approximately 1-1.5% kW reduction per degree reduction.
- Economic break even point starts around 50 tons.
- Water and waste costs compete with energy savings. Drain, fill, and freezing are design considerations.

This diagram shows the effect for air cooled equipment in a climate with design temperatures of 95 deg F dry bulb and 58 deg F wet bulb. Similar benefits are possible in different climates provided the wet bulb depression value is similar.

Adiabatic Humidification

Basis of savings: Evaporation without energy input to cause boiling.

- Includes evaporative pads, spray nozzles, atomization, and ultrasonic methods. Good results if combined with a process that simultaneously needs cooling. No energy advantage if the cooling effect is not beneficial and must be counteracted with new-energy heating.
- Psychrometric process follows the constant enthalpy line and increases moisture content as temperature drops.
- Compressed air and ultrasonic technologies each use approximately $1/10^{th}$ of the energy compared to boiling the same amount of water.

Note: all of the adiabatic evaporative methods require air at a reasonable temperature to humidify and generally will not work well in air temperatures lower than 50 deg F, therefore application is best suited to the return air stream or other tempered air stream and not in the outside air stream in cold weather.

Adjacent Air-cooled Equipment Too Closely Spaced

Basis of Savings: Correcting re-circulated discharge air condition lowers inlet air conditions so the air-cooled equipment energy use is lowered as if it were a cooler day.

- Warm air re-entrainment can occur from improper equipment spacing. Elevated intake cooling air directly raises refrigeration head pressure and compressor power by 1-1.5% per degree.

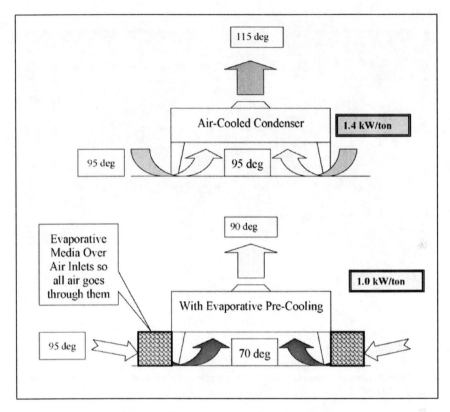

Figure 3B-2. Evaporative Pre-cooling

- A good rule of thumb for proper spacing is the air inlet or vertical finned coil height projected horizontally.

Constant Speed to Variable Speed Pumping Conversion
Basis of Savings: Pump energy use profile matching the heating/cooling load profile will use less energy than "constant volume" constant energy pumping.
- The key is to provide the fluid flow upon demand, but not all the time. For comfort systems, the system will be controlled to track the load profile. Applies to HVAC applications including chilled water pumping, condenser water pumping, and heating water pumping. Requires a variable speed controller for the pump and a load following signal.
- For example, if chiller load (tons) is known, the adjusted flow rate and speed for pumps can be derived automatically to proportionally

Figure 3B-3. Improper Air-Cooled Condenser Spacing
This installation was measured with 5 degrees of elevated inlet air temperature
and an estimated power increase of 5-7% as a result.

follow the load change. The most substantial changes will occur at hours when load is between 50% and 100% load (flow between 50-100%).

• Additional flow savings can be obtained by increasing system delta T (dT), which allows the thermal energy transport to occur with less mass flow and pump work.

• The savings of pump energy should be weighed against any change in compressor efficiency to be sure there is a net gain, especially from condenser water.

• Chilled water flows below 2 feet per second often encroach on laminar flow and can result in compressor energy penalties that negate pump savings. The prudent approach is to verify for the specific chiller the compressor kW/ton at the proposed changed flows and temperatures to assure there is a system benefit and not just a pump benefit.

Constant Volume to Variable Air Volume Conversion

Basis of Savings: Fan energy use profile matching the heating/cooling load profile will use less energy than "constant volume" constant energy air moving.

- Applies to any air moving system, including exhaust, supply, and make-up. Applies equally to process air movement as comfort systems. The key is to provide the air flow movement upon demand, but not all the time.
- The energy savings is from having fan energy track the load profile. Applies to comfort air conditioning and heating or any other air moving task with a variable load that can be served by varying the volume of air. Requires a variable speed controller for the fan and a load following signal, commonly a downstream pressure sufficiently high to allow VAV boxes to operate.
- For a given heat load (heating or cooling) and differential temperature, the required air flow can be easily calculated and is seen to directly follow the changing load. The most substantial changes will occur at hours when load is between 50% and 100% load (flow between 50-100%)
- Additional flow savings can be obtained by increasing system differential temperature (delta T), which allows the thermal energy transport to occur with less mass flow and fan work.

Constant Volume Terminal Reheat to VAV Reheat Conversion

Basis of Savings: Reduced overlapping heating/cooling and reduced fan horsepower.

- Each zone reheat coil is replaced with a VAV box. Most will require a heating coil unless serving only an interior area.
- Careful evaluation of the upstream duct system is required to be certain it will be suitable for the higher pressures involved.
- Heat is available only after the air flow has been reduced to "minimum cooling flow," therefore less reheating of supply air occurs.
- See the sketch of this modification on page 75.

Testing Adjusting and Balancing (TAB)

Basis of Savings: Reduce system pressure by excessive damper throttling, thereby reducing transmission energy requirements.

- This measure requires knowledge of existing positions of dampers or valves. In some cases, if operations staff or maintenance procedures over the years have spoiled the original balancing effort, the balancing status maybe a large unknown quantity and this measure may be as much about restoring confidence and performance as anything else. But, while it is being balanced, do so in a way that encourages low fan/pump energy input. Also, be sure to have locking provisions and permanent marks for the balancing devices, to help sustain the work.

- Impeller trimming for over-sized pumps can be a big energy saver, equal to the amount of energy dissipated at the balancing valve.

Proportional Balancing Method

Described for air, but is applicable to both water and air balancing.

- The system is first measured with all dampers fully open and the fan at full output.
- With the initial iteration, each outlet is measured and given a percentage of the design intended flow. Some may be above and some may be below the intended flow rates.
- The outlets are numbered in order of increasing percentage of design airflow. The outlet with the lowest percent remains open and is not adjusted. Its percentage is designated as B and is the basis for other branch adjustments.
- The flow rate on the branch with the next lowest percentage is adjusted so that it has the same percentage as B. All other branches are adjusted to this same value of B.
- When that initial step is complete, then adjust the main fan capacity (by dampers, sheaves, motor speed, motor change, etc.) to achieve full design capacity.
- Then return to the individual outlets and spot check at least 20% of them to assure that they are within the stated tolerance. If they are not, then repeat this process iteratively until they are. At the conclusion, there is still at least one damper that is fully open, to minimize overall system loss.

Remove Inlet Vanes or Discharge Dampers

Basis of Savings: Reduced System Pressure and Fan Hp.

After converting to VFDs, these devices create unwanted air pressure drop even in the wide open position.

The wide-open loss depends on the free area (FA) of the damper and the air velocity. Inlet vanes and control dampers are typically around 80% FA.

Inlet Vane and Outlet Damper Pressure Drop Loss
Source: American Warming & Ventilating
Based on the following relationship and free area factors.

$$(FA\ Factor)(Velocity/4005)^2$$
FA Factors
70% - (0.624)
75% - (0.434)
80% - (0.306)
85% - (0.224)

Lower VAV Duct Static Pressure Control Setting
Basis of Savings: Power is reduced exponentially with a reduction in pressure.
The relationship $HP2 = HP1 \times (^{SP2}/_{SP1})^{y1.5}$ is a derivation of the fan

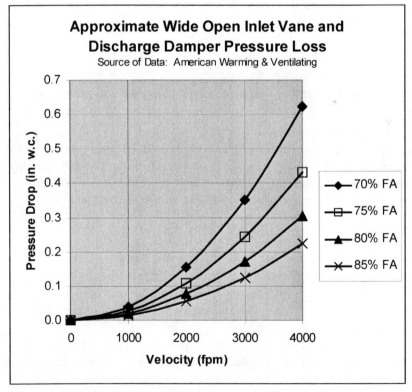

Figure 3B-4. Inlet Vane and Outlet Damper Pressure Drop Loss

laws that applies to VAV systems without the constraint of a maintained downstream duct pressure. However, most commercial VAV systems use a maintained downstream duct pressure and so this relationship must be de-rated. This de-rate is required since the pressure reduction is only occurring in the discharge duct section—losses in the balance of the system, including casing and coil losses, are not affected. So, only external losses contribute to this and of those, only the discharge section.

The exact value of the exponent depends on the proportions of internal to external duct resistance and the fraction of the external duct resistance impacted by the set point reduction.

Fraction of total air system resistance affected by set point reduction	Exponent
100%	1.5
75%	1.1
67%	**1.0**
50%	0.8
25%	0.4

Figure 3B-5. Fan Law Exponents for VAV Static Pressure Reduction

A short cut method that works for most standard VAV air handlers is to drop the exponent, such that the savings from reduced pressure varies proportionally with the pressure reduction. This assumes 2/3 of the fan energy is spent on the supply duct—which, by design is high velocity and restrictive and the most energy intensive portion of the duct system.

$$HP2 = HP1 \times (^{SP2}/_{SP1})$$

Reduce Resistance in Distribution Ducts and Pipes
Basis of Savings: Power is reduced with a reduction in resistance.

This model considers removing an obstruction or otherwise 'lowering the bar' of the whole system for pressure requirement and applies to both constant and variable flow.

Before considering changes to equipment and controls, consider the source and the distribution system. Begin by using less if possible: less air, less water. Do you need all that air or water circulating?

Possible ways to reduce system resistance:

Water
- Remove balancing valves after conversion to VFD
- Re-balance so at least one balancing valve is fully open
- Clean strainers earlier
- Replace restrictive main line strainers
- Replace restrictive main line flow meters (orifice plates)

Air
- Remove old inlet vanes or discharge dampers after conversion to VFD
- Re-balance so that at least one balancing damper is fully open
- Clean coils
- Change filters earlier
- Extended surface or angled filters
- Eliminate unnecessary duct appurtenances
- Smooth out 'bad' duct fittings

Power Reduction from Reduced Duct and Pipe Friction

Correcting Bad Duct Fittings, Entrance Losses, Exit Losses
Basis of Savings: Reducing system pressure reduces fan horsepower requirements.

The basic relationship for air horsepower is

$$HP(air) = CFM * \frac{TSP}{6356}$$

Where CFM is cubic feet per minute air flow and TSP is the total

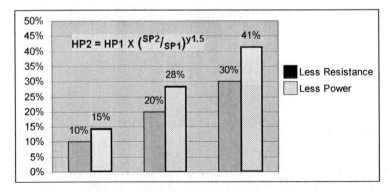

Figure 3B-6. Fan Law Exponents for VAV Static Pressure Reduction

static pressure, in. w.c.

Motor Hp is higher than air horsepower, because losses from fan efficiency, fan drive, and motor efficiency must be incorporated.

For a given air flow, reducing TSP directly reduces horsepower. In addition to the friction losses, coil and filter pressure drops etc., part of the work of the fan is fittings and entrance/exit losses. When one of these is especially bad, the pressure drop can be unusually high and an opportunity for improvement.

This book will not fully cover duct fitting losses. There are a number of excellent texts on this, namely *ASHRAE Fundamentals and SMACNA Duct System Design Manual.*

Some Rules of Thumb:
- Transitions should be no more than 30 degrees, and 15 degrees per side is better.
- Abrupt (blunt) changes in duct geometry are usually bad, unless velocity is very low (<1000 fpm)
- Square elbows should always have turning vanes
- Velocity is usually highest at the fan outlet
- Fan discharge conditions are high loss areas unless they have smooth transitions
- Losses increase exponentially with velocity
- Many bad duct fittings are there because of lack of room and don't have good solutions
- HVAC air velocity over 2500 fpm requires special care in design to avoid high losses

Losses are normally calculated in terms of "number of velocity heads" (Hv), where tables provide the "C" factor based on testing.

$$Hv \text{ (air)} = (V/4005)^2$$

where Hv is in. w.c., and V is velocity in feet per minute

Example:
A poor duct fitting has a C-factor of 0.9 and can be replaced with one having a C-factor of 0.20. What are the savings in hp if the air flow is 20,000 cfm and the duct velocity is 2000 fpm at that point?

Ans:

$Hv = (2000/4005)^2 = 0.25$ in. w.c.

Savings is the differences of "C," which is 0.9-0.20 = 0.7

Reduced pressure is 0.7 velocity heads, or 0.7 * 0.25 = 0.175 in. w.c.

$HP = (20,000 * 0.175)/6356 = 0.55$ hp

Often bad duct fittings are accompanied by noise, either from the fitting or from the fan laboring to move air through the turbulence.

Distributed Heating instead of District Heating
Basis of Savings: Reduced thermal losses at part load.

There are some good reasons to use district heating.

• Centralized maintenance
• Aesthetics
• Possible source of free heat source to utilize

However, energy of this system is usually higher than for a distributed heating system, where heating equipment is located closer to the point of use. The main reason for this is the thermal line losses. While these may be kept to a tolerable level at design loads, unless the heating fluid temperature is reset in mild weather, these losses are constant and form a larger and larger fraction of the total energy use at part load.

Figures 3B-7 and 3B-8 show overall system loss at part load, given full load thermal losses. These losses (2%, 5%, 10%, 20%) represent total distribution thermal losses (piping losses).

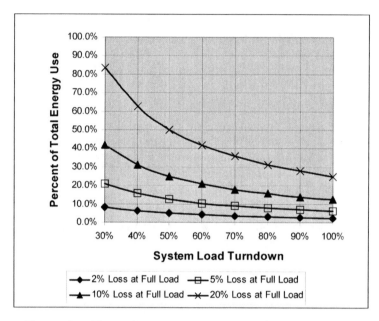

Figure 3B-7. Thermal Losses at Part Load (80% firing efficiency)

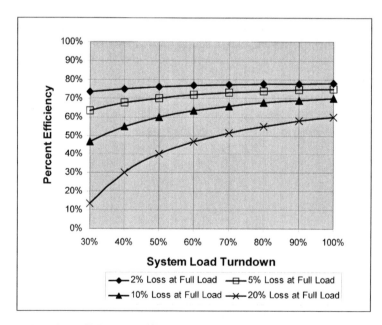

Figure 3B-8. Overall System Efficiency Loss at Part Load (80% firing efficiency)

ECM DESCRIPTIONS—SWIMMING POOLS

Pool Covers
Basis of savings: Isolates the water surface from air, preventing evaporation.

Reduce Pool Evaporation
Basis of savings: Water lost through evaporation absorbs heat, 1000 Btu per pound, cooling the remaining water. Additionally, the water lost is replaced and must be heated to pool temperature. Reducing evaporation reduces these losses directly.

Pool energy use is mostly due to water heating, and heating burden (other than initial heating) is largely due to evaporation, so controlling evaporation and other water losses is an important consideration in an energy program. Evaporation is strongly affected by two things: *wind speed* at the surface and the *differential vapor pressure* between the water and air, so energy use from pool evaporation can be reduced by:

- Lowering the pool temperature
- Raising the surrounding air temperature
- Raising the surrounding air humidity
- Lowering the wind speed at the air-water surface interface

Note: HVAC systems that exchange air to dehumidify should do so based on relative humidity controls, and excessive drying by using outside air will increase evaporation.

Recommended Pool Water and Air Temperatures
Surrounding air temperature should generally be higher than and within 2 degrees of the water temperature to reduce evaporation, although air temperatures over 85 def F are not recommended for comfort reasons. Higher humidity lowers evaporation rates, although humidity over 60% is not recommended to reduce risk of biological growth.

Figure 3B-9. Recommended Pool Air and Water Temperature

Type of Pool	Air Temperature, deg F	Water Temperature, deg F	Relative Humidity, pct rH
Recreational	75-85	75-85	50-60
Therapeutic	80-85	85-95	50-60
Competition	78-85	76-82	50-60
Diving	80-85	80-90	50-60
Whirlpool/Spa	80-85	97-104	50-60

Source: ASHRAE Applications Handbook, 1999, © American Society of Heating, Refrigerating and Air-Conditioning Engineers, Inc., www.ashrae.org.

Pool Evaporation Facts

Actual values will vary depending upon individual conditions, but the basic relationships can be illustrated as follows:

- At 50% rH, increasing wind speed over the water increases evaporation rate 30% for each 1-mph of wind speed increase.
- At 50% rH, reducing surrounding air temperature increases evaporation by 4% for each 1-degree lowered.
- At 60% rH, reducing air relative humidity over the water increases evaporation 2-3% for each 1-pct rH lowered.
- Mechanically heating the pool air over the water is 15% more efficient than using the pool to heat the surrounding air, due to reduced evaporation.

Pool Evaporation Formula

Simplified Pool Evaporation Formula
ASHRAE 1999 Applications Handbook

$$wp = 0.1A * (pw - pa) * Fa$$

wp	=	evaporation of water, lb/hr
A	=	area of pool surface, ft^2
pa	=	saturation pressure at room air dew point, in. Hg
pw	=	saturation vapor pressure taken at surface water temperature, in. Hg
Fa	=	Activity Factor:

	Residential pool:	0.5
	Condominium	0.65
	Therapy	0.65
	Hotel	0.8
	Public, schools	1.0
	Whirlpools, spas	1.0
	Wave pools, water slides	1.5

Figure 3B-10. Simplified Pool Evaporation Formula

ECM DESCRIPTIONS—HEAT RECOVERY

General Criteria

The term "heat recovery" is the commonly used term for a variety of systems that recovery a portion energy from a waste stream and put it to use, displacing new energy consumption that would otherwise occur. "Energy recovery" would be a more descriptive term since the concept applies to more than just heat. The most common application in industry is taking waste heat and using it to heat or pre-heat another mass that needs heating anyway, thereby saving fuel. The same principles apply for air or water flows needing to be cooled anyway—cooling or pre-cooling from waste streams that are cooler. Humidity recovery is also viable for large air streams by pre-humidifying from a more moist exhaust air stream, or de-humidifying make-up air with a drier exhaust air stream. In all cases, heat recovery serves to reduce new energy input by re-using waste products. While the goal is admirable, there are a number of practical barriers.

Heat Recovery Viability Test

For heat recovery to viable, it must meet these requirements:

1. *Same Time.* Generally, waste heat and the need for heat must occur concurrently. Thermal storage could assist this, although this is rare. To illustrate, drain waste pre-heating domestic hot water makes sense for a shower, but not for a bath.

2. *Right Proportions.* The only amount of recovered heat that counts is where the available heat and needed heat are equal. When there is an abundance of either waste heat or needed heat, it is the smaller of the two that determines the actual energy recovery effect. To completely eliminate the new-energy heating and supplant it entirely with waste heat, the waste heat quantity must exceed the need for heat recovered heat. Best economics occur when the two are in similar proportion.

3. *Right Temperatures.* Waste heat must be at a sufficiently higher temperature than the recovered heat sink to allow heat transfer. The greater the differential temperature available, the more economical the apparatus due to the higher "approach" temperature there is to work with.

4. *Enough Recovered Heat to Make it Pay.* There must be sufficient heat transfer throughout the year to provide economic justification for implementation. One rule of thumb is 5000 CFM of outside air intake with equal exhaust nearby. Single pass systems (outside air make-up, raw water heating) are good candidates since they are energy intensive. Facilities with a large number of operating hours per hear, especially continuous operations, provide quicker returns than others.

Heat Recovery Application Notes
*   New construction can benefit the most from heat recovery proposals since the cost of the recovery equipment is subsidized by the reduced size and cost of the primary heating equipment that would otherwise be provided. The down side to this is that the facility now becomes dependent upon the heat recovery system, and so its proper sizing, application and maintenance for sustained operation become more critical.
*   For retrofits the heating equipment is backup and the recovery benefit simply reduces its load. Savings in energy alone must pay for the recovery equipment and so returns are longer.
*   Recovered heat energy is penalized from additional air horsepower from pressure drop through coils, pumping energy, standby losses, and from capital investment requirements.
*   When either the waste heat or recovered heat temperatures vary over time, there can be times when heat transfer is marginal, or when heat recovery is detrimental. Whenever the differential temperature is near zero, the actual heat transfer will be very low and the energy used by active system components (fans, pumps) may not be justified. Controls monitoring the differential temperatures should be used to determine when to shut down the system.
*   For *refrigeration* system heat recovery, a good rule of thumb is 4000 Btuh per ton of refrigeration capacity available for hot water heat recovery.

Exhaust–to–Make up Air
Basis of savings: Reduced heating and cooling energy for make-up air, compared to 'raw' outside air.
*   Examples:
*   Gas clothes dryer vent used to pre-heat incoming combustion air.

- Paint booth exhaust pre-heating make-up air.
- Building exhaust pre-heating make-up air.

Rejected Heat-to-make-up Water
Basis of savings: Reduced heating energy for make-up water, compared to 'raw' make-up water.
- Examples:
- Boiler economizer, pre-heating make-up water.
- Refrigeration hot gas (before going to the condenser), pre-heating make-up water.

Rejected Heat-to-space Heat
Basis of savings: Reduced heating demand on the space heating system.
- Examples:
- Waste heat off an air compressor that can warm a section of a factory.
- Refrigeration system rejected heat used as heat in an air handler or as reheat in a dehumidification cycle.

Wastewater-to-make-up Water
Basis of savings: Reduced heating energy for make-up water, compared to 'raw' make-up water.
- Example:
- Commercial laundry waste water heat exchange to pre-heat wash water.
- Injection mold cooling jacket water to warm process water or boiler feed water.

Combined Heat and Cool: The Water-to-Water Heat Pump
Basis of savings: Compound savings: cooling energy saved and heating energy saved. Usually expressed as cheap heating and free cooling.
- Waste refrigeration heat can be used as a primary heat source, provided there is a concurrent need for both heating and cooling. The chiller pre-cools the chilled water return, shedding load on the main chillers, and the condenser becomes a water heater.
- Special equipment is needed to achieve refrigerant hot gas temperatures sufficient to make space heating water temperatures of 140 def F or higher. Coefficient of Performance values (COPs) go down at these temperatures but are still around COP=2.0 which is not great for a cooling machine but is very respectable for a heating machine.

Recovery of Humidified or De-humidified Air
Basis of savings: Where humidified or dehumidified exhaust occurs, the energy normally used for humidifying or dehumidifying the make-up air can be recovered.
- Special heat wheels with desiccant or other moisture holding material are used.

Examples:
- Building exhaust pre-heats and pre-humidifies outside air intake in winter.
- Building exhaust pre-cools and pre-dehumidifies outside air intake in summer.

Type	Efficiency (note 1)
Heat Wheel	60-80%
Heat Pipe	60-70%
Plate-Box	60-80%
Run-Around Coil	40-60%

Figure 3B-11. Air-side Heat Recovery Equipment Efficiency Guidelines

Note 1: thermal heat recovery potential. Does not include parasitic losses of fan or pumps, or added resistance of the recovery equipment in the fluid path. There are a number of variables that determine application efficiency such as relative temperatures and proportions of flow rates. Values vary by manufacturer—for example some low end commercial heat recovery wheels may be less than 50% efficient.

Double Use of Process Air and Water in Heat Recovery
Basis of savings: Both the energy and fluid itself are re-used directly, without heat exchange apparatus. The second point of use is seen as 'free'. Note that consideration of contaminants as well as implications of failure and shutdown of individual equipment is required.
- General exhaust from a theater used as make-up and cooling for a projector
- General exhaust from a building used as make-up for toilet exhaust or kitchen exhaust

- Steam condensate not being returned added directly to wash water
- Waste heat exhaust from an air compressor intercooler used for boiler make-up air in winter
- Single pass refrigeration cooler water used for cooling tower make-up

━━━━━━━━━━━━━━━━━━━━━━

ECM DESCRIPTIONS—THERMAL ENERGY STORAGE (TES)

Basis of savings: Savings for all thermal energy storage cooling systems is the ability, via storage, to use energy during off peak times when it costs less.

TES Pros and Cons

- Full storage systems are capable of keeping the chiller off the entire on-peak time. The storage systems cost more since the number of ton-hours is higher. However, conversions of existing conventional chiller plants to TES may be sufficiently sized and are candidates for full storage if the refrigeration equipment is in good condition with life remaining in the equipment.
- Partial storage TES systems serve to defray part of the on-peak demand and flatten the electrical load profile. They run concurrently with the refrigeration system during the day and run at night to re-charge the storage. This means the refrigeration equipment runs almost continuously. Still, these systems, especially on new construction, are less expensive to install.
- Flexibility is a key detriment to most of these systems. Even a properly sized system can be rendered obsolete if the rates change, and the chances of rates changing during a 20 year equipment life cycle are very good.
- Cool storage and warm storage systems are all plagued with stratification losses. Various attempts have been made to deal with it, and it remains an engineering challenge.
- Systems with cyclic freeze-thaw are often plagued with expansion damage. Design must include ample provision to accommodate these forces.

TES option comparison	Conventioal mechanical refrigeration	FULL STORAGE chilled brine ice storage	PARTIAL STORAGE chilled brine ice storage	FULL STORAGE chilled water cool storage	FULL storage evaporative cooling with cool storage
Max cooling load, tons	500	500	500	500	500
Annual load, ton-hrs/yr	720,000	720,000	720,000	720,000	720,000
Storage stand-by loss %	0%	5%	5%	20%	20%
% system storage	0%	100%	50%	100%	100%
Storage stand by losses	0	36,000	18,000	144,000	144,000
Total cooling load, ton-hours	720,000	756,000	738,000	864,000	864,000
Tons capacity	500	350	175	350	350
hours of storage	10	10	10	10	10
Design daily storage load, ton-hrs (tons*hrs*storage factor, incl storage loss)	0	5,250	2,625	6,000	6,000
Storage tank size, gal	N/A	95,000	47,500	600,000	600,000
Chjiller kW/ton	0.60	0.75	0.75	0.60	0.00
Auxiliary kW/ton	0.20	0.30	0.30	0.30	0.30
Total kW/ton	0.80	1.05	1.05	0.90	0.30
Total kW demand (kW/ton * installed tons)	400	368	184	315	105
Annual energy use, kWh (ton-hrs * kW/ton)	576,000	793,800	774,900	777,600	259,200
kWh on-peak	200,000	0	150,000	0	0
kWh off-peak	376,000	793,800	624,900	777,600	259,200
kW demand on-peak	400	0	184	0	0
kW demand off-peak	400	368	184	315	105
kW demand off peak above on peak	0	368	0	315	105
$/kWh on peak utility cost	0.06	0.06	0.06	0.06	0.06
$/kWh on peak utility cost	0.03	0.03	0.03	0.03	0.03
$/kW-yr on-peak utility cost	120	120	120	120	120
$/kW-yr off-peak utility cost	80	20	80	20	20
$ for kWh on-peak	12,000	0	9,000	0	0
$ for kWh off-peak	11,280	23,814	18,747	23,328	7,776
$ for demand on-peak	48,000	0	22,050	0	0
$ for demand off-peak	0	7,350	0	6,300	2,100
$ Total elec cost	71,280	31,164	49,797	29,628	9,876
M$ Total 20-yr elec cost	1.43	0.62	1.00	0.59	0.20
$ Maintenance cost per year	50,000	50,000	40,000	30,000	20,000
M$ 20-yr maintenance cost	1.00	1.00	0.80	0.60	0.40
M$ Installed cost premium	0.0	0.4	0.0	0.6	0.1
M$ Total 20-yr cost	2.43	2.02	1.80	1.79	0.70

Figure 3B-12. Sample TES Cost Comparison

Notes:
1. Parameters will vary by locale. This is intended to show how the factors to compare.
2. No first cost incentives considered.
3. For partial storage, any number of fractions of storage are possible. 50-50 was used for this example.
4. Cost of water considered equal for each option and not shown.
5. For full storage systems, a special thermal storage electric rate is assumed.

- For warm or cool storage, the container volume is an order of magnitude larger with correspondingly greater surface area (compared to ice), and stand-by losses become increasingly important. Standby losses are probably on the order of 20% for warm or cool storage.
- Minimum of 25% energy penalty in ice making mode, even with lower condensing temperatures at night.
- TES normally only makes economic sense when there is a large rate incentive for off-peak use, and never makes sense if energy conservation is important. If the rates are there, these systems can save utility bill money, but almost always use more energy, and almost always cost more to install.
- Significant off-peak utility rate discounts for energy and demand are usually required to make such systems attractive. However, other considerations may make thermal storage a good choice, such as ride-through back up for critical cooling applications, allowing hours of cooling in the event of a power loss while using generator power for circulating pumps only.

Rules of Thumb for TES Systems
- Cool storage: 100 gal/ton-hr (15 deg dT), 150 gal/ton-hr (10 deg dT)
- Encapsulated ice: 17-22 gal/ton-hr
- Ice on coil: 18-26 gal/ton-hr
- Installed TES system (ice), all types: approx $100 per ton-hour. Source: Cryogel, 2007.

Conditions Favoring Thermal Energy Storage
- Average cooling loads are much less than the peak cooling load.
- Large differential between on-peak and off-peak energy and demand charges.
- Low off-peak demand charges.
- High number of seasonal cooling (or heating) hours and ton-hours load.
- Utility incentives to defray first cost.
- Available space for storage containers.
- Higher-than-average operational staff technical expertise.

Cool Storage
- Chilled water is created during off peak times when power costs are less. The chilled water is stored in a large tank and used during the

day allowing the chillers to be turned off.

Ice Storage

- Ice is created with low temperature brine during off peak times when power costs are less. The chilled water is stored in a large tank and used during the day allowing the chillers to be turned off.

Phase Change Material (PCM) Storage

- Same as ice storage, except PCMs can be selected for phase change at temperatures closer to utilization temperature (45 def F) instead of ice (32 def F). This technology has the potential to leverage the compactness advantage of ice storage equipment without the 25% inherent energy penalty.
- The barrier is that PCMs are costly while water is not.

Cool Storage—Evaporative Cooling

- In dry climates, it is possible to use evaporative cooling at night in conjunction with cool storage, to reduce or eliminate mechanical refrigeration. This is a very good way to save energy in refrigeration—by turning it off completely—however pumping costs will be higher and will offset some of the savings. Cooling season wet bulb temperatures in the mid-40s and low 30s are needed to drive this. In dry climates, this variation has strong promise.

ECM DESCRIPTIONS—ELECTRICAL

Power Factor Correction

Basis of savings: There are two.

(1) Electric utility charges for power factor, and
(2) Decreased I^2R copper loss within the facility distribution wiring, from the wires carrying the excess magnetizing currents.
- Utility fees alone can often justify power factor correction projects.
- Example: For some utilities, the power factor charge is a 1-for-1 increase in demand charge. If the power factor is habitually 80 percent and the utility charges for anything lower than 95 percent, then the cost of poor power factor in this case would be (95-80) = 15

percent increased demand charges.
- Note: Simple capacitor installation may bring good results for facilities with mostly motor loads, welding, etc. But in facilities with high levels of harmonics should be studied very carefully and special and costly power factor correction equipment may be required.

Load Balancing

Basis of savings: Motor performance and efficiency presumes equal voltage on each phase. If the voltage and current are different, then one phase will pull harder than the rest, and the motor windings fight, with ensuing energy loss and motor heating.
- A 2 percent voltage imbalance on a polyphase motor can reduce efficiency by 5 percent. For example, a motor with an 85 pct eff. nameplate could be 0.85 * 0.95 = 80.75 percent efficient.

ECM DESCRIPTIONS—OTHER

Lower Compressed Air Pressure
Basis of savings: Reduces compression ratio and required compressor work.
- Approximate savings are 1% power for each 2 psi lowered.

Lower Compressed Air Inlet Temperature
Basis of savings: Increases inlet air density, increasing volumetric efficiency and required compressor work.
- Approximate savings are 1.9% power for each 10 deg F lowered.

Lower Steam Pressure
Basis of savings: Several Factors:
- Lower steam pressure means lower steam temperature, and heat equal heat transfer occurs at reduced combustion temperatures. Thus, reduced pressure reduces stack temperature and casing radiation losses.
- Reduces feed water pressure and pumping requirements
- Reduces losses from leaks
- Reduces condensate temperature and attendant losses

Lower Boiler Excess Air

Basis of savings: Reduces the cooling effect of the extra air, increasing combustion temperature and heat transfer. Excess air 'sweeps' heat out of the boiler with no benefit.

• Approximate 1% efficiency gain for each 15% reduction in excess air.

Pre-heat Combustion Air

Basis of savings: Increases combustion temperature and heat transfer.

• Approximate savings are 1% efficiency increase for each 40 deg F pre-heat.

Pre-heat Feed Water

Basis of savings: Reduces the cooling effect of the feed water on the boiler water, reducing heating load.

• Approximate savings are 1% efficiency increase for each 10 deg F of pre-heat.

PART C
The Energy Audit

The purpose of the energy audit is to systematically gather and analyze information in a way that helps identify waste and suggest opportunities for improvement. The energy audit is a tool to be used to understand and improve energy use.

This is a wise thing to do for many reasons: to save money or to control costs; to decrease maintenance requirements or to enhance reliability; or to avoid capital expenditures or perhaps to justify them.

Part C discusses the steps, tools, and methods to conduct an effective energy audit with checklists and samples of direct actions to take or recommendations to consider. It is among the first tasks necessary to accomplish an effective energy control program.

WHAT IS AN ENERGY AUDIT?

The energy audit process concerns itself with three simple things:

How energy is used
Cost of that energy
Actions to reduce

The goal is to save money. It is unfortunate that the term "audit" invokes apprehension at least as old as the Internal Revenue Service (IRS). While audits within the financial and security realms may often have ominous implications, this is not at all the case with energy audits.

An effective energy audit provides a systematic method to analyze energy use, propose ways to reduce consumption, and to evaluate the recommendations. Sometimes it is helpful to have a list of things to consider. Here is a list of some important components of an effective energy audit:

1. List of Audit Tools
2. Do your Homework First before you go out and Play (Audit)
3. Site Data Input Form (copy of Market Manager input form)
4. Breakdown of a Utility Rate Structure
5. Energy Balance

1.   List of Audit Tools

The first and most important tools in the auditor's toolbox may be personal safety devices and careful procedures. A good pair of safety shoes, hardhat, safety glasses and ear protection are fundamental to most situations. A digital camera helps to document visual elements.

Proper compliance to OSHA (Occupational Safety and Health Association) regulations should be observed at all times, with special attention to fall protection, asbestos awareness, confined spaces protocols, lockout/tagout procedures and a host of common sense precautions. In unfamiliar surroundings, always ask questions and remain alert and observant.

Electrically insulated gloves may be necessary if electrical measurements are being taken above 50 VAC. Other site conditions or circumstances may also warrant additional protection. For example, breathing masks may be required if fumes or hazardous materials are present.

A list of typical auditing tools are shown in Figure 3C-1. A tape measure, flashlight, and simple hand tools such as pliers and screwdrivers are often crucial to basic information collection and effective time utilization. Many audit procedures also involve voltage, current, temperature and light level readings.

Equipment nameplate data are often obscured or missing. A quality multi-meter can used for basic voltage and current measurements. These electronic meters typically provide digital voltage-ohm-ampere readings. An AC current clamp is a type of ammeter that measures current without disconnecting the wire conductor. It can be a self-contained unit or a probe attachment for the multi-meter. For both tools, when measuring the value of alternating current signals, models providing true root mean square values (true RMS) help avoid errors due to distorted waveforms or harmonics.

Thermometers and temperature probes allow for measurement of surface or process temperatures necessary to evaluate thermal status or heat recovery opportunities. Portable hand-held infrared thermometers are useful to collect preliminary readings quickly. These might sometimes be called laser thermometers because a laser may be used to help aim the device allowing a lens to focus infrared energy from a source on to a detector, which is electronically converted to display a unit of temperature.

A light meter with remote sensor accurate to about 3% of readings are affordable and provide convenient illumination data. Foot-candle readings taken at various locations in a facility can reveal areas that have improper light levels. A foot-candle can be thought of as the amount of light that actually falls on a given surface. Adequate lighting levels for specific tasks and workspaces are established by organizations such as the Illuminating Engineering Society (IES). Some areas may be using newer lighting technology but are wasting energy by providing excess illumination.

These are tools that cost in the few hundred dollar range. More expensive tools may be required for some applications. Infrared cameras, combustion analyzers, power meters, and data logger kits may all be employed in select applications to provide the necessary data to develop a complete picture of the facility.

2.   Do your Homework First, Before You Go Out and Play (Audit)

Understand the business environment and drivers of the customer sponsoring your audit. There are several considerations to help make your site visit as productive as possible. Much of this information can be gathered by interview or requested in advance of a site visit. For example:

(A)   Arrange to have people and documentation on site at your disposal; try to determine key people and decision makers. Answer the following questions:
- Who influences the decision to pay for and implement energy conservation measures (ECMs)?
- What do the influencers hope to accomplish besides a financial return?
- Who can operate and maintain the ECMs once implemented?
- Who within the customer organization is best suited to be a liaison for the technical energy audit and implementation?
- What are the customer facility safety requirements?

(B)   Assess the financial circumstances and business criteria. Answer the following questions:

Metering Tools
Power Meter with low power AC Current Clamp and Electrical Test Lead Set
Digital Multimeter
Thermocouple module for digital multimeter
Rotary air flow measuring module for digital multimeter
Infrared Thermometer
Digital Psychrometer
Micromanometer kit with pitot tube
Petes Plug temperature sensor (manual, visual read out)
Petes Plug temperature transducer for digital multimeter
Light Meter
Data Loggers - runtime
Data Loggers - process signals (4-20mA, temperature probes, etc.)
Data Loggers - with split-core A/C current sensors
Data Loggers - CO2 plus calibration kit
Tachometer (end driven RPM)
A/C Phasing Meter

Advanced Metering Tools
Ultrasonic listening device
Delta-Pressure metering device
Ultrasonic liquid flow meter
Air Hood
Infrared Camera

Other Tools
Digital camera
AA and AAA rechargeable battery and charger kit
Screwdriver set, 6-way screw driver, and allen keys,
Needle nose pliers, linemans pliers, and channel lock pliers
Cordless drill plus battery recharger
Flashlight
Receptacle plug wiring tester
Smoke Sticks (infiltration testing)

Safety Related Tools
AC Voltage Detector
Gear Box for Meters and Accessories
Insulating rubber gloves
Hard hat
Steel toed shoes
Protective Glasses
Ear Plugs
Kevlar gloves
First aid kit (in car)
Fire Extinguisher (in car)

Figure 3C-1. Audit Tools

- What sources of funding are available other than energy savings?
- What are the customer's financial critical success factors? (minimum SPB requirements)
- What are the customer's energy or commodity pricing constraints? (Do they have long term contracts?)

(C)   Determine contractual preferences. Answer the following questions:
- What is the customer's procurement method for the implementation of energy savings measures?
- Are there federal, state, or local codes that affect the implementation of ECMs, (e.g., FDA regulations for food processing plants)?
- Who owns the assets that will be modified as part of the ECMs?
- Any special construction specifications or requirements, or special commissioning requirements?

(D)   Assemble as much technical data as possible before going to the site including:
- Building Floor Area
- Age of Facility
- Site Drawings (especially handy are 8-1/2 x 11 fire escape plans)
- Utility Bills—Understand the energy use patterns, so that you can search for the causes of the anomalies while on site;
- Know as much as can remotely about the site to customize your auditing forms;
- Utilization: Occupancy and Schedule (including shutdowns, vacations, holidays, etc.)
- Zoning by Occupancy Types
- Description of Core Business
- Plug Loads (description)
- Process Loads, Process Equipment, Process Flow Diagrams, and Compressed Air Systems (description)
- List of HVAC, DHW, Lighting etc. Equipment (from Maintenance Work Order database or from Construction Drawings)
- % of Space Heated, % Cooled; Ventilated = Yes/No
- Automatic & Manual Control of Equipment (descriptions)
- Projects in Progress
- Projects in Planning
- Problems
- Wish List
- Past Renovations within last 5 years (age, description, handy to check utility bills before & after these renovations)
- Location of known Environmental Hazards (asbestos, etc.)

3.   **Site Data Input Form (Copy of Market Manager Input Form)**
Standardized input forms can expedite the acquisition of building data. Proper preliminary planning will identify requirements for which unique information or special measurement is required.

4.   **Breakdown of a Utility Rate Structure**
Utility bills have many components that differ by utility company

and tariff rates. Some elements can change each month. It is important to understand how the rate structure applies to energy costs at the facility being audited in order to properly calculate proposed savings and to analyze the utility profile.

5.   Energy Balance

The energy balance quantifies all the energy used or produced by a system. Inputs and outputs can be accounted for in this way and opportunities to reduce consumption may be identified.

6.   End-use Analysis

An energy profile establishes the amount of energy consumed for different purposes within a system or facility. Consumption and cost for a particular occupancy may be compared to benchmark data that helps to identify systems that may benefit from an upgrade or retrofit.

7.   Utility Bill Disaggregation by End–use

Disaggregate means to separate into parts to classify or analyze. In the sample for electrical demand in Exhibit 7a, the kW demand profile for eight separate items is shown over a 12-month period. Fossil fuel and electrical consumption (e.g., kWh) would be shown in a similar manner.

8.   Sample Walk-thru (Preliminary Energy Audit = PEA)
Report Outline

The format for the walk-through report provides an executive summary with some detail on energy consumption along with some retrofit recommendations for consideration. Specific energy conservation measures (ECMs) may be proposed for which some cost savings may be estimated.

9.   Sample Investment Grade Audit (IGA)
(Detailed Engineering Study = DES) Report Outline

A more complete detailed energy study includes a more thorough description of buildings, a detailed utility analysis, with specific recommendations for which cost avoidance estimates are made. Some ECMs may have been considered but not recommended. Operational benefits may also be identified. Methodology for measurement and verification is suggested as well.

10.   Sample M&V Plan Outline

Measurement and verification (M&V) is the process of using measurement to reliably determine actual savings created within an individual facility by an energy conservation measure (ECM). Savings

Figure 3C-2. Building Energy Simulation Input Form Used as a Site Audit Input Form

Market Manager Data Form
Data File Modified Date:

		Zone 1	Zone 2	Zone 3 CORE	Zone 4 EXTERIOR	Ref./Source
ZONE DATA	Total					
Conditioning		Htg/Clg/Ext	Htg/Clg/Ext	Htg/Clg/Ext		
No. of Stories						
Floor-Ceiling Ht. (Avg)						
No. of Occupants						
Roof Area						
Floor Area (Gross)						
Design Infil. Smr (ach)						
Design Temp Smr (F)						
Design Infil. Wtr (ach)						
Design Temp Wtr (F)						
Relative Humidity Smr. (%)						
Relative Humidity Wtr. (%)						
T-Stat Smr WkDay/setback						
T-Stat Smr Wkend						
T-Stat Wtr WkDay/setback						
T-Stat Wtr Wkend						
WALL CONSTRUCTION						
NORTH (wall)						
Area (ft^2)						
Height (ft)						
R-Value						
Mass (lb/ft^2)						
WINDOW						
Area (ft^2)						
R-Value						
Shading						
Tightness		loose/avg/tight	loose/avg/tight	loose/avg/tight		
Shading Coefficient						
SOUTH (wall)						
Area (ft^2)						
Height (ft)						
R-Value						
Mass (lb/ft^2)						
WINDOW						
Area (ft^2)						
R-Value						
Shading						
Tightness		loose/avg/tight	loose/avg/tight	loose/avg/tight		
Shading Coefficient						
EAST						
Area (ft^2)						
Height (ft)						
R-Value						
Mass (lb/ft2)						
WINDOW						
Area (ft^2)						
R-Value						
Shading						
Tightness		loose/avg/tight	loose/avg/tight	loose/avg/tight		
Shading Coefficient						
WEST						
Area (ft^2)						
Height (ft)						
R-Value						
Mass (lb/ft^2)						
WINDOW						
Area (ft^2)						
R-Value						
Shading						
Tightness		loose/avg/tight	loose/avg/tight	loose/avg/tight		
Shading Coefficient						
ROOF						
Exterior Color		D/M/L	D/M/L	D/M/L		
Ceiling Type		Bare/Suspend	Bare/Suspend	Bare/Suspend		
Area (ft^2)						
R-Value						
Roof Mass (lb/ft^2)						
FLOOR						

Figure 3C-2 (Cont'd). Building Energy Simulation Input Form Used as a Site Audit Input Form

Area (ft²) / Story						
Mass (lb/ft²)						
R-Value						
Exposed Perimeter (ft)						
OCCUPANCY						
Week Day Profile						
WeekEnd Profile						
USE DEFAULTS						
No. People / 1000	(ft²)					
Sensible Gain / Person	(Btuh)					
Latent Gain / Person	(Btuh)					
Installed Equip/Plug Load	(Watts/ft²)					
Installed Lighting	(Watts/ft²)					
Pk. Hotwater Usage	(gallons/day)					
Ventilation	(cfm/ft²)					
LIGHTING						
Lighting Input Cap.	(kW)					
Week Day Profile						
Weekend Day Profile						
EQUIPMENT						
Equipment Input Cap.	(kW)					
Week Day Profile						
Weekend Day Profile						
MOTORS						
	Horsepower					
Week Day Profile						
Weekend Day Profile						
SERVICE HOT WATER						
SHW Input Cap.	(kBtuh)					
Tank Temperature	(degrees F)					
	Fuel					
Week Day Profile						
Weekend Day Profile						
HEATING						
	SYSTEM					
	Min. Outside Air (%)					
	Suumer Start					
	WinterStart					
	Occupancy Begin					
	Occupancy End					
	Htg. Equipment					
	Htg. Input (kBtuh)					
	Htg. Output (kBtuh)					
	Auxiliary (kW)					
	Fuel					
	Economizer	Air/Enth None	Air/Enth None	Air/Enth None		
Supply	FAN					
	Capacity (kW)					
	Type	FC BI Axial	FC BI Axial	FC BI Axial		
	CFM					
Return	FAN					
	Capacity (kW)					
	Type	FC BI Axial	FC BI Axial	FC BI Axial		
	CFM					
COOLING						
	Equipment Type					
	Cooling Input (kW)					
	Cooling Output (Ton)					
	Auxiliary (kW)					
	(COP or kW/Ton)					
	Schedule (% on)					
Cooling Tower	FAN					
	Capacity (kW)					
	Static Pressure					
	CFM					

Notes:

cannot be directly measured, since they represent the absence of energy use. Instead, savings are determined by comparing measured use before and after implementation of a project, making appropriate adjustments for changes in conditions throughout the performance term.

M&V activities consist of some or all of the following:

- Meter installation calibration and maintenance
- Data gathering and screening
- Computation method and acceptable estimating techniques
- Computations with measured data
- Reporting, quality assurance, and external verification mechanisms

Using a combination of these activities, information and adjustments can help bring the data into conformity with all the variables, helping to normalize the savings. Normalizing these savings allows customers to compare the original baseline savings to the current period using a common set of conditions.

11. List of Energy Conservation Measures

The type and extent of ECMs that can be proposed as self funding improvements in today's marketplace is extensive and growing all the time. As energy costs escalate, implementing conservation strategies become critical to economic survival. Moreover, prospects for carbon offset strategies loom on the horizon.

12. List of Operational Measures

Reduced operation and maintenance activities represent distinct savings opportunities for some applications. It is important to document the source and amount of each of these cost reductions. Confer with the customer to validate all assumptions. If possible, cite explicit references in contract documents to avoid future confusion regarding the baseline criteria for each operational measure.

13. Milestones

As with any team endeavor communication is critical to success. Detailed milestones with indicated dates and responsible parties are instrumental in achieving a good outcome. Each energy audit has individual characteristics but written objectives with dates and named deliverables are invaluable to completing the project on schedule and within budget.

Figure 3C-3.

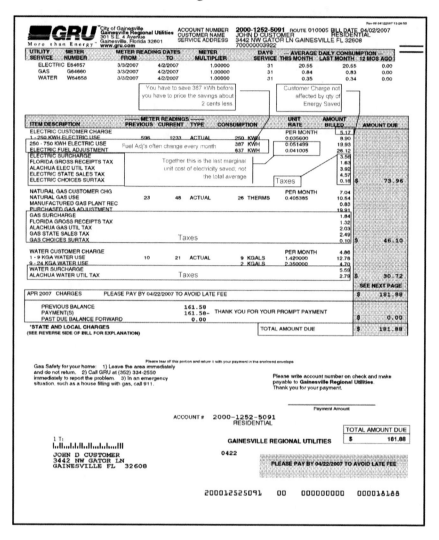

Figure 3C-4.

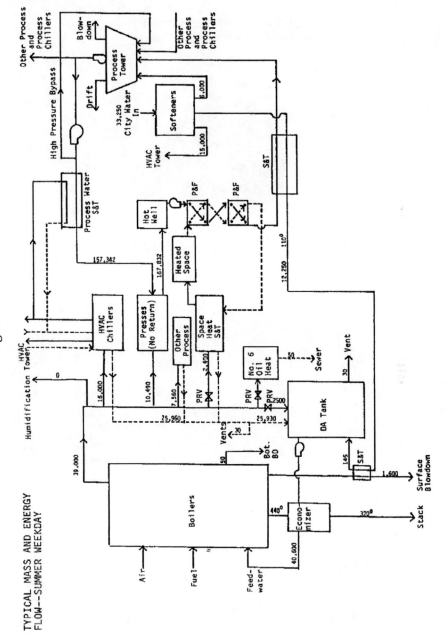

TYPICAL MASS AND ENERGY
FLOW--SUMMER WEEKDAY

Figure 3C-5.

ENERGY END-USE PROFILE
Somewhere High School - Fiscal Year 94-95

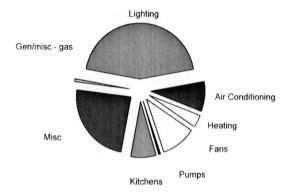

	Peak Demand					
End-Use	Summer (kW)	Winter (kW)	Annual Consumption		Annual Cost	Percent of Total
A. ELECTRICITY						
Lighting	70	70	115,841	kWh	$ 13,901	44.54%
Air Conditioning	123	5	22,404	kWh	$ 2,689	8.61%
Heating	1	8	9,100	kWh	$ 1,092	3.50%
Fans	16	15	26,612	kWh	$ 3,193	10.23%
Pumps	1.05	0.98	1,657	kWh	$ 199	0.64%
Kitchens	12	12	19,658	kWh	$ 2,359	7.56%
Misc	37	39	62,997	kWh	$ 7,560	24.22%
Subtotal	260	150	258,270	kWh	$ 30,992	99.3%
B. NATURAL GAS						
Gen/misc - gas			111	therms	$ 219	0.7%
Subtotal			111	therms	$ 219	0.7%
C. FUEL OIL						
Space Heating - oil			0	gallons	$ -	0.0%
Domestic Hot Water- oil			0	gallons	$ -	0.0%
Subtotal			0	gallons	$0	0.0%
C. TOTAL ENERGY COST					$ 31,211	100.0%

Figure 3C-6a.

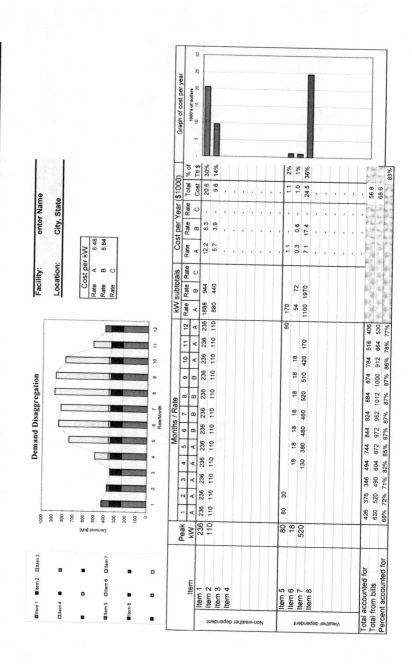

Figure 3C-6.b

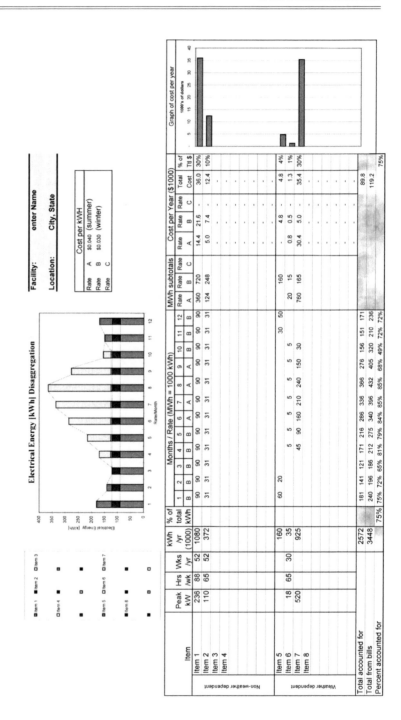

Figure 3C-6c.

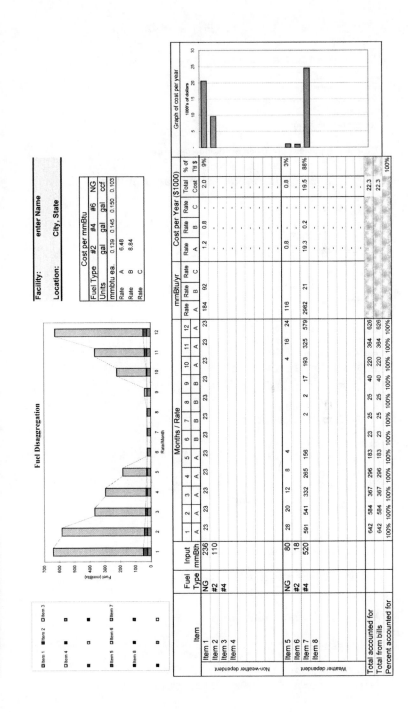

Figure 3C-7.

I.   Executive Summary

Program Objective
Preliminary Economic Opportunity
Next Steps
Report Organization

Customer support appreciation & acknowledgements

II.   Facility Description and energy profile

A. INTRODUCTION

B. FACILITY NAME #1
1. Facility/Systems Description
Lighting.
Space Conditioning & DHW.
Receptacle & Kitchen Load.
Control System.
Process Loads.
2. Energy Consumption and Costs
Electrical Consumption and Tariffs
Fuel Consumption and Tariffs
3. Previous Measures Implemented
4. Observations of Existing Conditions

C. FACILITY NAME #2
1. Facility/Systems Description
Lighting.
Space Conditioning & DHW.
Receptacle & Kitchen Load.
Control System.
Process Loads.
2. Energy Consumption and Costs
Electrical Consumption and Tariffs
Fuel Consumption and Tariffs
3. Observations of Existing Conditions
4. Observations of Existing Conditions

D. PREVIOUS MEASURES IMPLEMENTED

III.   Recommendations

A. INTRODUCTION

B. FACILITY NAME #1
Energy Conservation & Operational Improvement Measures
ECM #1

(Continued)

Figure 3C-7. (*Cont'd*)

ECM #2
ECM #3

C. FACILITY NAME #2
Energy Conservation & Operational Improvement Measures
ECM #1
ECM #2
ECM #3

D. GENERAL FACILITY AREASFACILITY NAME #2
Recommendation #1
Recommendation #2
Recommendation #3

Figure 3C-8.

1. Executive Summary
A. Program Objective
B. Preliminary Economic Opportunity
C. Next Steps
D. Report Organization

2. Facility Description and Energy Profile
A. Introduction
B. Facility/Systems Descriptions
C. Utilities and Tariffs and End-Use Analysis and Load Dissaggregation
D. Previous Measures Implemented
E. Observations of Existing Conditions

3. Recommendations
A. Introduction (provide caveat that following ECM descriptions are generic)
B. Energy Conservation and Operational Improvement Measures (ECMs & OIMs)

ECM Title
a. Description/Rationale
b. Existing Conditions
c. Proposed Conditions
d. Energy Savings Calculation / Methodology
e. O&M Impact
f. Outline Specification
g. Cost Estimate Parameters (what will make this the expensive version or the cheap version)
h. Monitoring Plan (M&V lite)

C. Measures Considered But Not Recommended (McBnr's)

4. Scope of Work for Next Steps
A. IGA Work Plan (work required to detail generic ECMs to site specific conditions - if this is a preliminary Energy & Ops Assessment)
B. Energy Services Agreement (if this was an IGA)

5. Appendices
A. Letter of Intent
B. Exhibits
C. Survey Records

Figure 3C-9.

M&V Plan for ECM /ecm name-description/ : The following describes the Measurement and
Verification procedures, formulas, and stipulated values which may be used in the determination of cost
avoidance and/or performance against the Guarantee.

1.ECM General Description & MV-Option Definition and Rationale

2.Boundary - of Energy Use and Savings - Determination

3.Baseline
3.1. Baseline – Conditions & Energy/Water Data
3.2. Planned and Unplanned (Routine and Non-Routine) Baseline Adjustments

4.Potential-to-Generate Savings Verification Plan

5.Parameters to be Monitored & Sampling Plan
5.1. Proposal (IGA, TA/EA) Period
5.2. Pre-Installation & Baseline Verification
5.3. Installation & Acceptance Period
5.4. Performance Period (On-Going)

6.Determination of Energy Avoidance & Determination of Cost Avoidance

7.Determination of Non-Energy O&M Cost Impact

8.Customer Responsibilities and ESCo-Customer Risk Allocation
8.1. Customer Responsibilities
8.2. Risk Allocations

9.Performance Period Definition and On-Going Activities
9.1. Performance Period Definition
9.2. On-going Activities

10.Reports & Documentation

Figure 3C-10.

A. Building Envelope
1. Install double/triple glazing.
2. Install storm windows.
3. Reduce glass area (wall up/close off).
4. Install solar shading, screening, blinds, awnings.
5. Replace clear glass with reflective glass.
6. Install insulated doors.
7. install airlocks, vestibules.
8. Wall up/close off unneeded openings.
9. Install Insulation (roof, ceiling, wall, floor).

B. Heating
1. Replace inefficient burners.
2. Replace inefficient boilers.
3. Downsize system.
4. Install automatic flue dampers.
5. Replace pilot lights with electronic Ignition.
6. Preheat combustion air/make up water with waste heat.
7. Recover waste heat from exhaust air, flue gas, laundry, kitchen, engine exhaust, condenser, cooling tower.

C. Cooling
1. Replace inefficient chillers.
2. Install window air conditioners for local load requirements.
3. Install economizer cycles.
4. Utilize evaporative/dehumidification cooling.
5. Manifold chillers in parallel and sequence.
6. Insulate off—line chillers and cooling towers.
7. Replace air cooled condensers with cooling towers.

D. Ventilation / Distribution
1. Reduce air volume.
2. Reduce air stratification.
3. Convert to variable air volume,
4. Insulate pipes and duct work.
5. Install automatic dampers.
6. Consider zoning mo4ifications.
7. Reduce outside air percentage.
8. Shutoff/reduce heat to lobbies, stairwells, hallways.
9. Reduce/eliminate air to unoccupied areas.
10. Utilize outside air for free cooling.
11. Eliminate simultaneous heating and cooling.

E. Lighting
1. Convert incandescent- to fluorescent.
2. Convert fluorescent to high intensity
3. Convert mercury vapor to metal halide or sodium vapor.
4. Modify fixtures (add reflectors1 lower height).
5. Reduce number of futures/lighting levels.
6. Delamp and disconnect ballasts.
7. Disconnect unused ballasts.
8. Employ task lighting.
9. Install occupancy sensors (infrared, ultrasonic).
10. Install local switches.
11. Install skylights.

F. Domestic Hot Water
1. Install flow restrictors.
2. Install self—shutoff—faucets.
3. Decentralize hot water heating.
4. Add piping and tank insulation.
5. Electrically trace heat supply piping to eliminate return piping and pumps.
6. Install separate hot water generator for summer
7. Install boost heaters for very hot water in lieu of central system use.
8. Add preheat from waste heat source

G. Laundry
1. Install heat reclamation system on laundry wash water. -
2. Install heat reclamation system on dryers.
3. Shut off equipment and appliances whenever possible.
4. Install makeup air supply for exhaust.
5. Install local boost heater.
6. Add/Improve insulation.

H. Kitchen
l. Shut off range hood exhaust whenever possible.
2. Install high—efficiency steam control valves.
3. Shut off equipment and appliances whenever possible. -
4. Install makeup air supply for exhaust.
5. Install heat reclamation system for exhaust heat.
6. Turn off lights in coolers.
7. Install nighttime automatic steam cut off.
8. Utilize chemical dishwashing system.
9. Add/improve Insulation.

I. Utility Plant Systems
1. Reduce steam distribution pressure.
2. Increase boiler efficiency.
3. Insulate boiler and boiler piping.
4. Install economizers.
5. Install air preheaters.
6. Install blow down controls.
7. Modernize boiler and chiller controls.
8. Replace steam traps.

J. Electrical Equipment
1. De-energize equipment when not in use.
2. Reduce loads when not required.
3. Install capacitors or synchronous motors to increase power factor.
4. Reduce transformer losses by proper loading and balancing.
5. Convert to energy efficient rotors.
6. Instill two speed / variable speed motors.
7. Replace oversized rotors.

K. Controls
1. Install automated energy management system.
2. Instill temperature/pressure reset devices.
3. Install stop/start devices.
4. Install night setback devices.
5. Install load shedding devices.
6. Install system optimizing capability.
7. Install enthalpy controls.
8. Replace hand valves with automatic.

I.. Renewables
1. Install active or passive solar hot water system.
2. Install active, or passive solar space conditioning.
3. Utilize photovoltaics.
4. Convert to wood.
5. Convert to biogas.
6. Convert to refuse.
7. Utilize wind energy.
8. Utilize water power.
9. Utilize geothermal.
10. utilize tidal energy.

K. Miscellaneous
l. Install cogeneration system.
2. Install thermal storage system.

SOURCE: private correspondence, but have been told that it might match a HPAC publication

Figure 3C-11.

A. Building Envelope
I. Repair door and window caulking.
2. Repair door and window weatherstripping.
3. Replace broken glass.
4. Adjust door closer.
5. Petit doors and windows.
6. Install temporary storm windows.
7. Turn off decorative fountains.
8. Install air curtains.
9. Cover roof ventilators when not in use.

B. Heating and Cooling
1. Lock thermostats.
2. Adjust supply or heat transfer medium temperature.
3. Restrict heat and air conditioning in unoccupied areas.
4. Clean radiators and air registers.
5. Clean condenser coils and fins.
6. Check operation of automatic controls.
7. Reduce heat in garages, docks, and platform areas.
8. Balance heating system.
9. Evaluate humidification system.
10. Check operation of all electric heating units.
11. Establish a regular program to inspect, clean, and lubricate equipment.
12. Lower-(winter) or raise (sinner) indoor thermostats.
13. Lower condenser water temperature.
14. Increase evaporator water temperature.
15. Defrost evaporator coils regularly.
16. shut off ice makers when unneeded.
17. Repair or replace damaged or missing insulation.
18. Isolate unused areas and rooms.
19. Isolate areas having extended occupancy.
20. Reduce occupied hours if possible.
21. Reschedule off—hour activities to accommodate partial shutdown of building systems.
22. Close interior shading devices (repair/replace).
23. Repair dampers and controllers.
24. Repair leaks.
25. Maintain cooling tower water quality.
26. Shut off unneeded recirculation pumps.

C. Ventilation / Distribution
1. Shut down system for unoccupied areas.
2. Improve operation controls (tine clocks).
3. Reduce ventilation rates to code minimums.
4. Reduce outside air intake to code minimums.
5. Repair dampers and controllers.
6. Balance air intake to occupant load.
7. Balance intake and exhaust air rate.
8. Improve mechanical operation (fans, motors, dampers).
9. Clean/replace air filters.
10. Maintain positive interior pressure.
11. Inspect all central systems and unitary controls.
I?. Discontinue use of unnecessary exhaust fans.
13. Replace/adjust worn/slipping fan belts.

D. Lighting
I. Reduce illumination levels.
2. Maximize use of daylight.
3. Use higher efficiency lamps.
4. Reduce or eliminate evening cleaning.
S. Clean lamps, diffusers, walls, windows, skylights.
6. Improve reflectance of surfaces.
7. Utilize task lighting.
6. Use lower wattage lamps.
9. Reduce or eliminate decorative lighting.
10. Discontinue use of nultt—level lamps.
11. Relocate high illumination tasks adjacent to windows.
12. Replace yellowed lenses with lenses that do not discolor.
13. Reduce outdoor lighting where practical.
14. Ensure outside lights off during daylight hours.
15. Train personnel to turn out lights, even for short periods.
16. Mark ganged switches to indicate areas controlled.
17. Repaint with light colored, non-glossy paint.

E. Domestic Hot Water
1. Repair leaks.
2. Reduce the quantity of water used.
3. Lower hot water temperature setting.
4. Check efficiency of oil or gas-fired equipment.
5. Repair insulation on pipes and storage tanks.
6. Clean heat pipe evaporator coil.
7. Shut off recirc ptacs when not required.
8. Repack pump packing glands to reduce leakage.

F. Laundry
1. Develop efficient operating procedures.
2. Iron only items which require it.
3. Wash and dry full loads only.
4. Reschedule work hours.
5. Switch to cold water wash.
6. Clean lint filters regularly.
7. Shut off equipment whenever possible.

G. Kitchen
1. Cook with lids on pots and kettles.
2. Preheat ovens only for baked goods.
3. Reduce temp or turn off fry tables and coffee urns during off-peak periods.
4. keep ovens and fryers loaded.
5. Use dishwasher for full loads only.
6. Maintain refrigerator door seals.
7. Shut off equipment, appliances, and range hood exhaust fans whenever possible.

K. Utility Plant Systems
I. Adjust barometric dampers.
2. Monitor boiler makeup water.
3. Operate minimum number of boilers.
4. Isolate off line boilers.
5. Check condensate return system.
6. Repair boiler, tank, and pipe insulation.
7. Check boiler efficiency and monitor combustion.
8. Eliminate gas pilots. -
9. Reduce stew pressure.
10. Repair faulty radiator shut off valves.
11. Check operation of steam traps.
12. Repair all leaks.
13. Clean plant and distribution system equipment.
14. Adjust air/fuel ratios.
15. Optimize scheduling and use of elevators.
16. Encourage people to walk I floor up; 2 floors down
17. Turn off vending machines, food warmers, coffee pots during unoccupied periods.
18. Turn off refrigerated drinking fountains.
19. Raise refrigerated drinking water temperature.
20. Disconnect transformers when serving no load.
21. Maintain transformer temperature at proper level.
22. Schedule use of major equipment to avoid peak electrical demand periods.
23. Remove soot and scale from boilers.

SOURCE: private correspondence, but have been told that it might match a HPAC publication

Figure 3C-12.

- Milestone 1—Confirm Program Goals, Objectives and Challenges
- Milestone 2—Determine Required Facility and Process Use
- Milestone 3—Identify Energy & Water Consuming Devices
- Milestone 4—Estimate Electrical Energy Use
- Milestone 5—Estimate Fuel Energy Usage
- Milestone 6—Estimate Water Use.
- Milestone 7—Measuring Actual Energy Use
- Milestone 8—Quantify Savings Potential
- Milestone 9—Delivery of the Final Proposal

Chapter 4

How to Implement a Solar Photovoltaic Project

Paul B. Breslow, Ph.D.
Senior Renewable Energy Analyst—Suntech Energy Solutions, Inc.

INTRODUCTION

The terawatts of solar energy available to the Earth in one minute exceed the energy demand for the entire world for a year. If even a fraction of this free, clean energy could be captured and converted to electricity, the need to convert expensive, non-renewable fossil fuels to electricity could be dramatically reduced.

Solar energy can refer to a variety of technologies that have been developed for each of the three general market categories—residential, commercial (which includes businesses, non-profits, institutions, and governments), and utility scale solar—based upon the amount of energy that the technology produces. The technologies can be generalized into the two broad categories of solar electric and solar thermal. Solar electric includes the various photovoltaic (PV) devices and systems that use the photovoltaic effect to convert light into electricity. Solar thermal includes products and technologies that range from the typical, black four foot by eight foot residential hot water heating panels to commercial and utility scale water heating that can be used for manufacturing processes, heating ventilation and air-conditioning (HVAC), and even power generation. Commercial scale solar PV projects generally refer to systems between 10 kilowatts (kW) and 2 megawatts (MW), since this is the range of sizes that can offset a high percentage of the usage (kilowatt hours or kWh) in most buildings while still being permitted by utilities and regulators. In addition to using PV, utility scale solar projects can use concentrating solar power (CSP) technologies developed by companies like eSolar Inc., which use reflective surfaces to heat water (or another fluid) up to the temperature that can be used for power plant operation.

To many people, solar photovoltaic systems conjure up images of unattractive, expensive PV modules that require batteries to provide the only source of electricity for a rustic cabin far from civilization (typically a 0.5 to 2 kW system). While this type of off-grid PV system is still common for remote, small power needs, the largest segment of the current PV market is for grid-tied residential, commercial, and utility scale power systems—and many of these systems can be financed over time at a cost similar to or even less than current utility rates. Additionally, today's aesthetically pleasing, dual-use PV options include shingles, tiles, skylights, windows, carports and facades that can be low maintenance, power producing, inexpensive alternatives to standard building materials such as glass and marble exteriors. Whereas off-grid PV systems provide the only source of power for those customers (and typically must use batteries for nighttime operation), many individuals and organizations are making the choice to own or host a PV system in order to offset their "brown power" usage with the clean, "green energy" produced from the PV. Brown power refers to electricity produced from fuels such as coal, natural gas, oil, and nuclear, which have some emissions or (potential) pollution associated with them. Conversely, green power refers to electricity and energy generated from non-polluting, renewable sources such as solar, wind, geothermal, ocean power, and some biomass technologies.

Photovoltaic power is rapidly gaining ground as the costs of hardware and installation come down, incentives from government and utilities are applied, and the costs of fossil fuels rise. Since PV systems are very low maintenance and produce power during the peak demand and price periods (especially during the middle of hot summer days when critical peak power pricing can reach $0.10 to $1.00/kWh or more), PV projects can often result in net annual savings or only a minimal premium over utility power. Environmental concerns are another powerful driver of solar's growing acceptance. Solar installations provide visionary customers the exciting opportunity to produce their own power, offset part of their greenhouse gas emissions, enhance their organization's reputation with a "green" statement affirming their commitment to the quality of the environment, and hedge against rising utility rates by fixing a portion of their electricity bill—often with savings from day one.

SOLAR PV 101

Solar PV modules and systems are rated by the number of Watts (W) that they produce in direct current (DC) or alternating current (AC). Since most residential and commercial scale buildings and devices require AC, all grid-tied systems require an inverter to convert the power from DC to AC. The three PV ratings are: Watts peak DC STC (W-DC-STC or Wp), which refers to the manufacturer's nameplate DC rating under standard test conditions (STC); Watts DC PTC (W-DC-PTC), which refers to the PV USA test conditions (PTC) estimate of real world performance DC; and Watts AC CEC (W-AC-CEC or W-AC-PTC or Wac), which refers to expected power output from the system and into the customer facility using the California Consumer Energy Center (CEC) ratings for PV equipment. A typical solar photovoltaic (PV) module is an aluminum framed, glass coated panel that is approximately 3.5 ft by 5.5 ft (~20 square feet) and produces 100 to 300 Watts peak (Wp) of DC power. A typical residential system might have six 200 Wp modules (1.2 kWp, approximately 120 sqft.) that would produce just over a 1 kilowatt hour AC (kWh) in direct sunlight and 1000 to 2000 kWh per year (depending upon location and system configuration), after accounting for the inverter and other system losses (typically ~ 15-20%). Similarly, a typical commercial scale PV system might have five hundred 200 Wp modules (100 kWp, ~85 kWac), which would require approximately 10,000 sqft and generate 100,000 to 200,000 kWh/yr. Utility scale solar PV projects range from 2 MW to 100 MW or more and connect directly into the utility's power lines.

Grid-connected solar PV systems do not require a choice between solar or the utility company. One gets the always-on reliability of the utility plus the clean energy from solar because the system is designed to work in parallel with the utility grid. The following major components are shown the schematic above.

Solar PV Modules—PV modules convert sunlight into electricity via the photovoltaic process.

Inverters—These and other power conditioning equipment convert the direct current electricity from the modules to utility-grade alternating current.

Meter—Meters measure system production and net energy usage, turning backwards (and crediting one's electric bill) when the solar system generates more energy than is needed on-site.

Utility Grid—The grid supplies supplemental power when demand

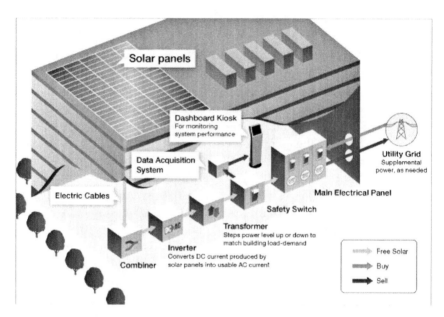

Figure 4-1. Grid-connected PV System (Suntech Energy Solutions, Inc.)

exceeds the solar PV system's output and absorbs excess when system output exceeds on-site demand.

The customer's electrical load draws power from either the solar PV system or the utility grid seamlessly and transparently. For safety reasons, grid-tied PV systems usually disconnect (stop producing power) during power outages to prevent power from being transmitted to the grid while utility workers are repairing the lines. For those who need a power backup solution, batteries or another type of energy storage can assist (depending upon the facility needs, the system design, and the net-metering rules).

THE BENEFITS OF SOLAR PV

Solar energy offers many distinct benefits over utility power and other forms of on-site electrical generation. While the most obvious benefits of solar are the utility cost savings, there are many other valuable benefits—financial, environmental, and societal (the three components of the triple bottom line measure of organizational success).

Financial Benefits

Solar Can Be Free or Cash Flow Neutral

In locations with good combinations of higher incentives, insolation, and utility rates, the solar PV project savings (and tax benefits) can be greater than the financing costs. Additionally, solar power purchase agreements (PPAs), where the customer "hosts" a PV system owned by a third party financier and pays only for the power produced, can often match or beat the avoided cost of power ($/kWh).

Current Incentives Can Make Solar Cost Competitive

Federal, state and local agencies provide a range of market building incentives and tax breaks to accelerate the payback on solar PV investments. Once the payback is achieved, the cost of electricity generated from solar energy is nearly free. The incentives are designed to help build the market for solar to scale, thereby driving costs down. Therefore, as costs come down, the incentives will be phased out and project economics will theoretically be similar. In reality, temporary market disequilibria will benefit some customers more than others—with the general rule that those customers who are first to sign up for a new incentive program typically receive the greatest incentives and best project economics.

Reduces the Highest Cost Electricity

Since PV produces during the peak price periods of the day and year (in time-of-use, TOU markets), solar helps customers to avoid the highest cost power.

Protection Against Rising Electricity & Fuel Costs (Solar Hedge)

An important advantage of solar energy over fossil fuel based utility and on-site power is that there are almost no variable costs—no fuel costs and minimal operating costs. Therefore, the cost of electricity generated by these systems is fixed for the life of the system, providing a valuable hedge against electricity price volatility and increases in utility rates.

Reduced Cooling Costs

Depending on your facility's roof deck construction and color, a roof-mounted solar PV system will shade your roof and reduce daytime surface temperatures up to 30°F. This can reduce heat transfer through the roof and the cooling load on the building by as much as 25%.

Reduced Roof Maintenance Costs

Due to roof shading by the PV system, the sections of roof under the modules will be subjected to less UV degradation and thermal cycling, thereby extending roof life.

Avoided Transformer Loss Costs

Facilities that have transformers on-site can have the solar power routed directly into the internal grid at the lower operational lower voltage, thereby eliminating the losses associated with stepping down the voltage from the transmission lines.

Add to Facility Value

The 30+ year potential for power production and any secondary PV system benefits, such as solar carports or canopies, add value to your facility.

Brand & Marketing Value

While it is difficult to quantify the dollar value of the increase in brand recognition and perceptions of corporate goodwill due to installing a solar PV system, there are obvious benefits. In addition to the press releases and commissioning ceremony, many organizations and their employees gain community recognition for being leaders in sustainability and environmental awareness. Customers are increasingly showing a preference for companies with sustainable practices.

High Reliability & Low Maintenance Costs

Systems have a design lifetime of 30+ years, and the module manufacturer warrants the solar panels for 20-25+ years. Most solar energy systems have no moving parts (other than inverter cooling fans), so routine maintenance is limited to visual inspections, inverter checks, and rinsing the modules with water occasionally to remove accumulated dust.

Renewable Energy Credits (RECs)

Each kilowatt-hour of solar electricity that a system generates contains two valuable products. The first is the electricity itself, which is identical to the brown power, e.g., the electricity with resultant pollutants that is generated from fossil fuels. The second is the "green" attributes; e.g., the pollutants that are not released into the atmosphere. The renewable attributes are measured in the form of "green tags" or "renewable energy credits," or RECS (and other terms). For every megawatt hour (MWh) of solar power (equal to 1,000 kWh), the system owner accrues one REC, which may be retained ("retired") or sold. Each REC produced offsets on the order of 400-2,400 lbs of carbon dioxide (CO_2), depending upon the generation fuel mix, thereby allowing the REC owners to claim the carbon offsets. These green attributes are often highly valued in the marketplace, particularly in locations where local utilities are required to derive a certain percentage of their total generation from renewable sources. Organizations that retain the RECs may claim that they are "solar or green powered." Organizations may choose to sell the RECs in order to offset the system costs, thereby capitalizing on the value of the RECs they generate as markets develop in parts of the US and around the world.

Environmental Benefits
Solar Power Produced is 100% Pollution Free

Solar energy is one of the cleanest forms of energy you can buy. By using solar energy, you reduce your organization's share of pollution (such as mercury, nitrogen oxides (NO_x), sulfur oxides (SO_x), etc.) caused by fossil-fuel powered generators and make a positive impact on global warming. While nuclear power does not generate air pollution, the radioactive toxic waste generated is difficult to safely secure, transport, and store.

Carbon Dioxide & Global Warming

Since solar energy does not involve any type of combustion, there is no carbon dioxide (CO_2) produced from PV plant operation (although there is a minimal amount of emissions associated with PV manufacturing, which is far, far less than the CO_2 associated with power plant construction and operation). CO_2 is one of the top contributors to global warming, therefore, solar PV helps to reduce global warming.

Energy Payback

The embedded energy (the total energy, including resource mining, required to manufacture) in PV modules only takes approximately one year to payback, thereby producing about 30 times the amount of energy required to manufacture it.

Societal Benefits
Silent Operation

Contrary to loud gas turbines or diesel generators, PV systems are virtually silent.

Removable & Recyclable
    In the event that the PV system needs to be moved, it can be uninstalled and re-installed in a new location. At the end of its useful life, it can be remanufactured into something useful.

Solar Carports & Canopies
    These provide customers and employees with much appreciated protection from the sun (and rain).

Reduced Demand for New Power Plant Construction
    Since PV systems are small, distributed power plants that produce during the periods of greatest demand (generally), building solar power systems reduces the need to build new power plants—especially natural gas or oil peaker plants that only run in the afternoon when the air conditioning load is high.

Reduced Dependence on Imported Energy
    Solar and other renewable energy projects require no fuel, thereby helping to reduce reliance on imported energy.

Economic Development
    Renewable energy projects have been shown to create more local (temporary and permanent) jobs than central station plants, since the bulk of the lifetime power cost is in the manufacturing and installation (not fuel).

Increased Pride of Ownership & Affiliation
    Studies have shown that organizations that demonstrate leadership by implementing green technologies generate more pride in those organizations' employees and customers, which leads to increased appreciation and decreased costs (i.e. recruiting, PR, permitting, etc.)

COMMON SOLAR PV CONCERNS

    Some of the potential concerns that people have about solar PV include:

Cost
    The initial investment in solar can be significant; however, incentives and new financing options such as the solar Power Purchase Agreement can make solar cost neutral or even reduce today's energy bills.

Space Required

While it is true that a PV system may require sizable square footage to deliver a significant amount of energy, one can usually utilize unused rooftop space or elevated designs to install a sizeable system.

Daylight Dependent Operation

Solar PV only produces during daylight hours (in direct proportion to the intensity of the sunlight), which dictates that nighttime energy supplies must be drawn from the standard grid or from another energy generation or storage system; however, since grid-connected systems are not relying solely on this power and the PV is offsetting the highest cost power, this is not an issue for most customers.

Decreased Performance Due to Climate and Shading Obstructions

Geographical location and local obstacles such as buildings, trees, etc., can affect the optimal performance of a solar installation; therefore, most solar integrators will design the system to be minimally shaded (or completely unshaded). Additionally, while solar insolation (sunshine) varies significantly across the US, almost any location in the continental US has a higher solar resource than Germany, the largest PV market in the world.

Technology Obsolescence

A common concern is that any technology purchased today will be obsolete in the near future, and that solar will be cheaper and more efficient next year. While technologies will incrementally improve and there will be breakthroughs in R&D, it takes years to develop the manufacturing processes necessary to produce mass quantities of high quality, robust, highly efficient products, with demonstrated field experience that are able to be credibly warranted by the manufacturer. Over the last five decades, solar has slowly decreased in cost and increased in efficiency—and this trend will continue.

Weight & Roof Penetrations

For rooftop systems, it is common—and appropriate—for customers to be concerned about the potential for a PV system to damage the roof or void the warranty. Qualified integrators (those with applicable experience and professional processes) use or suggest that the original roofing contractor (or a qualified consultant) verify the suitability of the roof for PV and then certify the installation in order to protect the existing roof warranty. Alternatively, some system types do not require penetrations and some building integrated photovoltaics, ("BIPVs") become an additional roofing layer.

AN OVERVIEW OF THE SOLAR PROJECT PROCESS

Implementing a commercial scale (10 kW to 2 MW) solar PV project can be broken down into the following seven steps.

Assess
An initial feasibility study by a qualified integrator or consultant will determine if PV will make sense from a design and/or financial perspective.

Design
A preliminary design can be developed after a physical site visit, which will allow the integrator to develop sufficiently accurate pricing to generate a proposal or bid.

Purchase or Financing & Rebate Application
Once the system installation price and cash purchase economics are known, the integrator or customer can request quotes for a PPA, lease, or loan and apply for any rebates.

Engineering & Permitting
Once all contracts are signed, the initial design can be finalized, the engineering drawings can be prepared, and the appropriate building permits can be pulled.

Procurement
Once a final design is developed, the integrator can procure the actual modules, inverters, mounting system, and other PV specific components can be sourced.

Construction, Commissioning, & Rebate Request
PV system installation can be quite straight forward and qualified integrators will minimize impacts to customer operations.

Operation and Maintenance Servicing
A web-based monitoring system (often required for incentives) and detailed O&M agreement (or plan) will ensure optimal system performance and life.

STEP ONE: ASSESS

Can solar power work for you? Even before initial consultations with a solar integrator, one can investigate the prospects by visiting some Internet sites such as the U.S. Department of Energy's National Renewable Energy Laboratory to learn the basics about solar photovoltaic systems and other technologies, such as concentrating solar power systems, passive solar heating and daylighting, solar hot water, solar process heat, and space heating and cooling.

If one engages an integrator, they should quickly provide a detailed assessment of the estimated power production, potential obstacles on or around the facility, the costs of at least one system configuration, and an overview of the financial incentives available. All these vary by location, utility, and state, so the information at this stage must be focused on the client's exact parameters and requirements, including the customer's financial situation and financing preferences. The decision should take into account the possibility that energy rates will rise in the future, so the merits of a long-term hedge against future costs should be weighed, as well as changing incentives, plans to own or occupy the location, and options such as buying, leasing, and power purchase agreements (PPAs).

This stage should also include an energy analysis that details how, when, where and why you use energy—most importantly, each meter's usage and demand. This audit will help set expectations for the system size that is appropriate to the usage, and moreover, it may reveal potential energy and dollar savings that can be captured by other means such as equipment retrofits, rate schedule changes, and participation in demand response programs offered by the utility.

Where Solar Projects Make Economic Sense

The first objection to solar is usually cost. Therefore, it is important to first ask if it has to make economic sense, and, if so, what hurdle it must reach. While this can measured in terms of payback, internal rate of return (IRR), net present value (NPV), or return on investment (ROI), there are many benefits—such as brand enhancement, publicity, and financial hedging—that may be difficult to accurately value. External factors such as increased carbon compliance legislation, interruption of gas supply, and competitive pressures will create additional risks for organizations that use large amounts of fossil fuels.

The economics of a PV project are dependent upon three factors—

insolation, incentives, and utility tariffs. The best economics will be in locations that have the highest combination these three factors. Understanding the engineering, legal (incentive program rules), and financial details and assumptions necessary to accurately model the project economics is difficult, but larger solar integrators should have qualified staff who can perform and explain this analysis for customers.

Insolation

Insolation is a measure of the amount of solar radiation (Watts) that falls on a specific PV surface area in a given time period. The insolation (or the production factor metric) of a solar PV array encapsulates both the geographic location and the system configuration, and is usually expressed in kWh/kWp (and understood to be per year). For example, a 100 kWp (100,000 Wp, i.e. 500 200Wp modules) system with a 1,500 kWh/kWp production factor would produce 150,000 kWh per year. The charts from the National Renewable Energy Laboratory (NREL) shown here demonstrate the relative production factors of a single axis tracking system versus a fixed tilt system at latitude minus 15 degrees.

Incentives

The broad category of solar PV incentives refers the various federal, state, utility and non-profit programs that make solar more cost effective. A good solar integrator will assist a customer in identifying, understanding, and applying for the applicable incentive types—and ensuring that they can be optimally used. The legislation that provides the rebates and tax incentives is often temporary, which can lead to quick program over subscription and boom-bust cycles of renewable energy development. The important aspects of incentives are:

Net-metering and Interconnection Rules

Most state net electricity metering (NEM) rules require utilities to purchase power generated from solar PV systems at the same price the utilities would otherwise charge the clients (termed full retail net metering, since some NEM programs only credit customer at the utility's marginal production or fuel cost). If the system is producing energy in excess of what it is using, the energy is sent into the utility grid and the building power meter actually runs backwards. Since many buildings are closed two days a week, this often means customers are routinely selling power back to (or banking kWh credits with) the utility. Interconnection rules regulate how large of a system that a facility is allowed to interconnect to

Figure 4-2. Fixed Tilt Insolation (NREL)

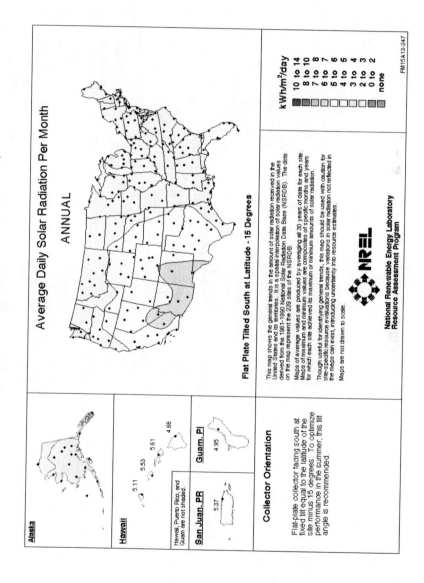

Average Daily Solar Radiation Per Month

ANNUAL

Flat Plate Tilted South at Latitude - 15 Degrees

This map shows the general trends in the amount of solar radiation received in the United States and its territories. It is a spatial interpolation of solar radiation values derived from the 1961-1990 National Solar Radiation Data Base (NSRDB). The dots on the map represent the 239 sites of the NSRDB.

Maps of average values are produced by averaging all 30 years of data for each site. Maps of maximum and minimum values are composites of specific months and years for which each site achieved its maximum or minimum amounts of solar radiation.

Though useful for identifying general trends, this map should be used with caution for site-specific resource evaluations because variations in solar radiation not reflected in the maps can exist, introducing uncertainty into resource estimates.

Maps are not drawn to scale.

kWh/m²/day

- 10 to 14
- 8 to 10
- 7 to 8
- 6 to 7
- 5 to 6
- 4 to 5
- 3 to 4
- 2 to 3
- 0 to 2
- none

FM15A13-247

Alaska

Hawaii

5.11 5.53 5.61 4.66

Hawaii, Puerto Rico, and Guam are not shaded.

San Juan, PR   Guam, PI

5.37   4.95

Collector Orientation

Flat-plate collector facing south at fixed tilt equal to the latitude of the site minus 15 degrees. To optimize performance in the summer, this tilt angle is recommended.

NREL

National Renewable Energy Laboratory
Resource Assessment Program

Figure 4-3. Single-Axis Tracking Insolation (NREL)

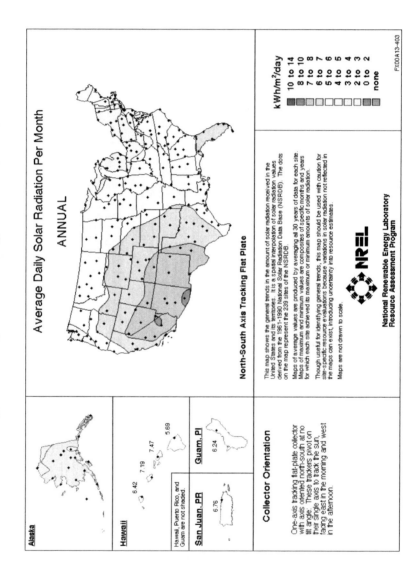

the grid. These NEM and interconnection rules are often, but not always the same. The map from the Database of State Incentives for Renewable Energy (DSIRE) shown here demonstrates which states allow net-metering and the maximum system size (note that Public Utility Commissions sometimes can be successfully petitioned to make exceptions).

State and Utility Rebates

Rebates are based upon either the size or performance of the installed system. Capacity based incentives (CBI) are in dollars per peak DC Watt or AC Watt ($/Wp or $/W-AC-CEC). Performance based incentives (PBI) are in dollars per kWh produced ($/kWh) over a specified term. There is general equivalence between the two methods (where a financial discount factor such as 8% is used to determine the other). Often, smaller projects (under 100 kW-AC-CEC) are paid an upfront PBI rebate called an expected performance based buydown (EPBB). In renewable energy credit (REC) compliance markets (states that allow utilities to purchase RECs to satisfy their renewable portfolio standard, or RPS), such as New Jersey, or in states with an RPS solar carve out, such as Colorado, solar RECs (SORECs or S-RECs) are sold by the owner/developer to finance the system (prices can range from $0.05 - $0.30+/kWh for up to 20 years). Feed-in tariff incentives, which pay customers directly for the power that their PV systems produce and put onto the grid, are common in Europe, but not currently common in the US (customers continue to accrue normal utility charges, but the power sold from the PV system provides a revenue stream). The DSIRE map here shows which states offer incentive programs as of March 2008 (though the value varies dramatically).

Tax Benefits

Tax benefits at both the state and federal level are a critical component of a solar project. They can be subdivided into four categories:

- *Tax Credits*—available from federal and state governments.
- *Accelerated Depreciation*—these depreciation schedules (such as the 5 year Federal Modified Accelerated Cost Recovery System, or MACRS) can often provide a post-tax value of 15-25% of the project cost. While rebates are more common at the state level, some states offer corporate income tax credits and accelerated depreciation, as well.
- *Sales Tax Exemptions*—Many states provide sales tax exemptions for PV systems.

Figure 4-4. Net Metering Rules by State (DSIRE, October 2008)

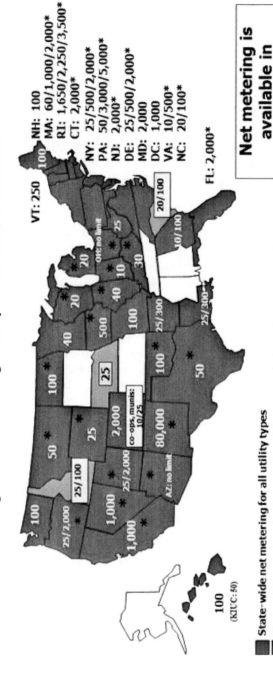

Net metering is available in 44 states + D.C.

VT: 250   NH: 100
           MA: 60/1,000/2,000*
           RI: 1,650/2,250/3,500*
           CT: 2,000*
           NY: 25/500/2,000*
           PA: 50/3,000/5,000*
           NJ: 2,000*
           DE: 25/500/2,000*
           MD: 2,000
           DC: 1,000
           VA: 10/500*
           NC: 20/100*

FL: 2,000*

State-wide net metering for all utility types

State-wide net metering for certain utility types only (e.g., investor-owned utilities)

Net metering offered voluntarily by one or more individual utilities

Note: Numbers indicate individual system size limit in kilowatts (kW). Some states' limits vary by customer type, technology and/or system application; this is the case when multiple numbers appear for one state. Other limits may also apply. For complete details, see www.dsireusa.org.

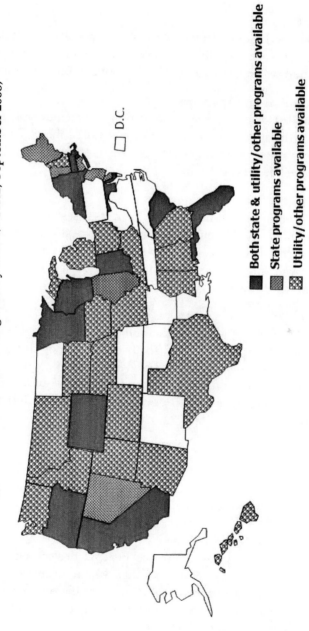

Figure 4-5. Solar Incentive Programs by State (DSIRE, September 2008)

■ Both state & utility/other programs available
▨ State programs available
▨ Utility/other programs available

□ D.C.

- *Property Tax Exemptions*—Many states provide property tax exemptions for PV systems.

Incentive Caps

Rebate and tax incentives are usually capped at some dollar value or percentage of system cost (or size), so it is important to understand if there is an optimal system size that will lead to optimal project economics.

Rebate Application Paperwork

Qualified solar integrators (some programs require integrator training and certification) will be able to provide all of the paperwork necessary to ensure that customers receive their incentives, and they usually will manage this process for the client.

Utility Tariffs

Utility tariffs differ not only by the rates assessed for each bill component, but also by the temporal (peak vs. off-peak) and seasonal (summer vs. winter) definitions. The map shown here is from the 2006 US DOE Electric Power Annual Report, and it shows the relative costs of commercial power across the US.

Tariff Structure

While most people refer to their utility rates as a single $/kWh number, non-residential electric bills usually are broken out into a number of components that are based upon the energy usage (kWh) and maximum demand (kW). Additionally, customers with a time-of-use (TOU) tariff will have these charges broken down into each period (often, off-peak, part-peak, and peak).

Solar Friendly Tariffs

Tariffs that include higher energy (kWh) based charges during peak solar production hours (daytime) and lower or non-existent demand (kW) charges lead to higher values for the average avoided cost of solar power.

Average Avoided Cost of Solar Power

This is the average estimated cost per kWh of electricity produced by the PV system that the customer would save if a specific solar PV system were installed. It is calculated by summing up the solar benefits, such as the TOU value of the energy produced and any estimate of demand savings, and dividing by the total kWh produced.

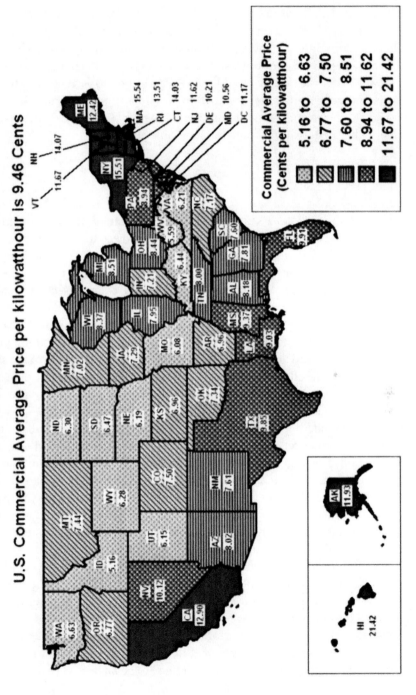

Figure 4-6. Average Commercial Utility Rates 2006 (DOE)

Utility Rate Escalation

This is often an area of contention, as there are many sources of data that measure different energy indicators over different time periods. Utility rates have risen over the last 30 years, and they will continue to rise—and at a faster rate, many would argue. In addition to increasing raw fuel costs and generating asset installation costs, local, national, and international policies are driving energy prices higher. The US Department of Energy has historical rate data for states and utilities.

STEP TWO: DESIGN

Installing a solar PV system is a relatively straightforward residential, commercial, or industrial construction project that is complicated by the fact that the installer is building a power plant. This challenge is mostly won or lost in the initial design, analysis, and engineering phases. Quality of design is an imperative for a successful solar project. This stage of a solar project needs to take into account not only the design of the installation itself, but the context in which it needs to work. Functionality and aesthetics need to be given equal consideration. The design may undergo revisions and adjustments in order to address particulars of the site, such as: the presence of construction equipment; landscaping; physical obstructions; existing electrical equipment; repairs, if required, to prepare roof mountings; etc.

System shading during any portion of the day or year will reduce system performance; however, there is a tradeoff between system size and production. The largest system will produce the absolute most, but it may not be quite as efficient as a smaller system, e.g. a smaller unshaded system that is optimally tilted will have a higher production factor (be more efficient from a whole system perspective) than a larger system that is less optimally tilted or sometimes shaded. Minor differences in system production factors can have significant effects on project economics.

There is no real off-the-shelf solar success, so the design process is essential for fitting your solar installation to your unique needs and the site. It may be here that you decide not to proceed. Candid communications with your integrator are vital to your decision. This is the stage at which the integrator should perform a full-site engineering audit to detect possible problems before they reveal themselves at a later and more expensive time. Questions arise here that need to be answered here: Is there adequate space to meet your goals for power-generation? At what angle should panels be pitched? If the installation is roof-mounted, is the roof in good repair? How will the installation affect the aesthetics of the site?

The solar PV industry has seen record breaking investment and growth since the turn of the century, and especially since the US Energy Act of 2005 instituted the 30% federal investment tax credit. However, the basic technology, such as the efficiency of PV modules and system configurations, has not evolved significantly in the last few years.

PV Technology

The solar cells inside the PV modules are the heart of the whole PV system. There are a variety of semiconductor materials and manufacturing processes used, but they share the basic concept of stringing together PV cells and—except in the case of some BIPV products—covering them with glass and encasing them in aluminum.

Crystalline Silicon

Monocrystalline (cSi) and multicrystalline silicon (mcSi) solar PV is the predominant technology that has been steadily improving over the last 50 years. Typical crystalline silicon modules efficiencies range from 12% to 20% and have warranties from 20 to 25 years.

Thin Film

this is a broad range of technologies that is gaining market interest. Thin film technologies are those that have a very thin layer of semiconductor material deposited on a thin substrate. They are typically referred to by their chemical composition, and the most common ones are copper indium gallium selenide (CIGS), cadmium telluride (CdTe), and amorphous silicon (aSi). Efficiencies are typically 4-10% and warranties are 20-25 years. Some reports show that thin film modules degrade at a faster rate than crystalline modules, but since they have not been field tested for as long, we must rely upon manufacturer claims and warranties.

BIPV

Building integrated PV materials are roofing shingles, roofing membranes, windows, skylights, and awnings that are both an integral part of a building and producing electricity. BIPV products utilize either crystalline or thin film technology (or a combination) encased in a common product form factor (such as a roofing shingle).

Mounting Type & Location

While PV systems have historically been designed for rooftops to minimize the allocation of otherwise useful space, the desire for larger systems and ancillary benefits are driving innovative design configurations. PV customers commonly consider the following three system locations in order to achieve dual use from the space and maximize system size.

Rooftop

Still the most common system built, rooftop mounted systems are fairly inexpensive. PV can be installed on any type of roof, but some roofing types, such as standing seam, are easier than others. Design considerations include penetrations or ballasting that is suggested or mandated by wind and seismic codes and the age of the roof. When considering installing a PV system that may operate in that location for 10-40 years, it is important to consider the type, age, warranty, and general condition of the roof.

Ground-mounted

Ground-mounted systems are typical in areas where land is not as much of a premium, where the system host can allocate the property for 10-40 years, for larger systems, and for typical single axis tracking systems. While this mounting type can often be the cheapest for large systems, the geotechnical requirements due to the results of soil analysis, the grading of the terrain, and the distance (for trenching) to the meter are the primary sources of increased cost.

Carport or Canopy

These types of systems produce power above a parking lot or garage and have the added benefit of shading the cars underneath (or covering a picnic area, etc.). This mounting type has been developed into many designs, with the more aesthetic designs and installations in more challenging locations leading to higher costs. Due to the necessary size of the steel, aluminum or concrete structure, carports are typically the most expensive mounting type. A key source of value in a PV project can often be the inclusion of carport/canopy costs (and associated tax benefits).

System Tilt, Orientation, and Tracking Method

PV modules are typically installed at a fixed tilt from zero to latitude plus 15 degrees and can be fixed or on some type of tracker. A general rule of thumb for maximizing the annual kWh production of fixed tilt systems

Figure 4-7. Rooftop PV Systems at Google (Suntech Energy Solutions, Inc.)

Figure 4-8. Ground Mounted PV System (Suntech Energy Solutions, Inc.)

in the northern hemisphere is latitude minus 15 degrees. The diagrams shown illustrate tilt, azimuth, and tracking actions.

Tilt

A typical fixed tilt system would be tilted at 0 to 45 degrees. A fixed tilt array configuration is defined by the tilt and azimuth of the array(s).

Figure 4-9. Carport PV Systems at Google (Suntech Energy Solutions, Inc.)

Azimuth

The optimal azimuth is near 180 degrees (due south). In order to optimize the space on a roof, systems are often designed to be orthogonal to the rooftop. Therefore, if the building is not facing directly due south, the PV array often will not be facing due south.

Tracking

There are a variety of tracking methods, but the most common are single axis (1x) and dual axis (2x) trackers. Trackers allow PV modules to be oriented as close to perpendicular to the sun's rays as possible, which

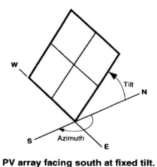

PV array facing south at fixed tilt.

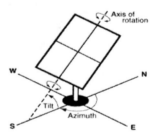

One axis tracking PV array with axis oriented south.

Figure 4-10. PV Array Facing South at Fixed Tilt (NREL)

Figure 4-11. Single Axis Tracking PV Array with South Azimuth (NREL)

optimizes the power produced by the modules. Single axis trackers that rotate from east to west about a north-south axis are the most common and cost effective. Trackers can sometimes be tilted to optimize the tracking for certain latitudes.

Since many rebate programs are structured to incentivize solar customers to configure systems in a certain way (i.e. optimized for summer production or a certain tilt), it is important to have the solar integrator explain their proposed design, calculate the exact incentive that the system will be eligible for, and demonstrate that it is optimal from an economic perspective (if that is the overriding criterion). However, it is often the case that important design constraints such as building orientation, shading, and the desire to maximize system size (i.e., lowering tilt to increase system size) will result in a slightly lower performance-based incentive, which results in the optimal system for the customer overall.

STEP THREE: PURCHASE OR FINANCING & REBATE APPLICATION

Solar electricity can involve no capital sacrifice, and can often be cash flow neutral or even result in utility bill savings. The most important single decision in the solar project process is deciding how to finance the system—either internally or with some type of external financing. Incentives from some state rebate programs, federal tax credits, and other tax incentives can reduce over 90% of the system cost. However, since the bulk of the incentives are tax based, non-profit organizations (including governments) and companies with insufficient tax appetite will be unable to receive all of the value of (monetize) these benefits if they purchase a system. Previously, companies that filed as alternative minimum tax (AMT) could not use the federal 30% ITC, but as of October 2008 that is no longer the case.

Therefore, a variety of third-party ownership financing options (such as power purchase agreements and tax advantaged leasing) have been applied to the commercial solar market that allow a customer to host a PV system and enjoy the benefits without the upfront cost. While a direct purchase allows businesses to retain all the benefits and project returns, many companies and non-profits opt for the power purchase agreement (PPA) or tax lease structure because they cannot fully monetize all the tax incentives or would rather invest in their core operations. Additionally, many organizations and project champions find it easier to gain management approval for a relatively small operating expense (or savings) rather than a larger capital expense with unfamiliar financial assumptions.

Since a PV system may be in place for 10 to 30+ years, it is important that the facility is owned or under a long-term lease. Financiers will require lease-terms at least equal to their finance terms. The four major financing options are:

Direct Cash Purchase

The customer uses cash on hand to invest in the solar project. This gives customers the opportunity to realize the full monetary benefits of the project and enjoy immediate system ownership.

Debt/Loan/Capital or Finance Lease

The customer uses the solar PV system and/or the real property as collateral to secure a debt structure on the PV system. This allows system ownership from day one with a known buyout (purchase) cost and, depending upon the project economics (due to incentives, system production, avoided electricity costs, tax situation, and interest rate) can lead to a cash flow positive project.

Tax-advantaged Leasing

The customer's credit rating and long-term facility lease or ownership permits a third-party financier to finance the system construction, leaving the customer to operate and maintain the system. While leases are more familiar than PPAs, PPAs often have lower annual costs and have no production or O&M risk, which make a PPA more appealing than a lease. For tax reasons, the buyout cannot occur in less than six years and must be the greater of fair market value (FMV) or a set price.

Power Purchase Agreement (PPA)

The customer's credit rating and long-term facility lease or ownership permits a third-party financier to finance the PV system construction, maintenance, and operation. There are a variety of PPA pricing models (or solar services agreement types), but the most common, simple, and financeable is a set start rate ($/kWh) that escalates from 0-6% depending upon the terms and customer's comfort with higher escalators (i.e. a $0.12/kWh start rate and a 3.5% escalator). Similar to tax leases, the buyout cannot occur in less than six years and must be the greater of fair market value (FMV) or a set price. Solar PPA customers pay for solar power, not for solar equipment or installation, thereby reducing the risk and O&M responsibilities. Being long-term agreements, PPAs are well suited to customers who own their own site and plan long-term occupancy. This financing structure was introduced in 2005 and is growing to become the most popular way for businesses, governments, and non-profits to go solar.

The table on the following page summarizes the various financing options, the major differences, and responsibilities.

In addition to the financing mechanisms listed above, there are many federal, state, and local programs—such as grants, low (or even zero) interest loans or bonds (i.e. federal Community Renewable Energy Bonds)—that can help finance or buydown the system cost. Some issues to be aware of when deciding between financing options include the following:

Insurance

The PV systems should, and depending upon the financing type must, be insured. Sometimes this can be under the customer's existing umbrella policy, but other times the financier will require and/or provide (and charge for) separate insurance (there could be tax/ownership implications for whomever insures the system).

RECs

Since RECs have a value (usually $0.005-$0.05/kWh for 1 to 20 years in voluntary markets vs. $0.05-$0.30+/kWh for up to 20 years in compliance markets), it is important to decide if your organization wants to retain the RECs for green or speculative purposes, or if it is worth it to give up the RECs in order to lower the PPA or lease rate offered by the financier. Organizations that wish to claim the green power or carbon offsets should retain and "retire" the RECs. Those interested in the speculative value of RECs (as opposed to those strictly interested in the lowest rate possible) may wish to retain the RECs or negotiate an option to purchase them at a specified price at any point in the contract. The premium that the financier places on the value of RECs and the alternative cost of purchasing RECs or carbon offsets from a green power or carbon marketer may dictate a particular course of action.

O&M

As mentioned above, some financing packages will include O&M and others will not. It is important to know who is responsible and what

PV Project Financing Methods	Direct Cash Purchase	Loan, Debt, Bond, Capital/Finance Lease	Tax (Operating) Lease	Power Purchase Agreement (PPA)
Who Owns the System?	Customer	Customer	Financier	Financier
Who Receives the Rebate?	Customer	Customer	Financier	Financier
Who Receives Tax Benefits?	Customer	Customer	Financier	Financier
Up Front Cash Requirement	Full Amount	None to Full Amount	None (usually)	None (usually)
Capitalized or Expensed	Capitalized	Capitalized	Expensed	Expensed
Rent Payments	None	High	Medium	Low
Total Economic Benefit Captured by Customer	High	High	Low	Low
Responsible for Maintenance	Customer	Customer	Customer	Financier
Performance Risk	Customer	Customer	Customer	Financier
Termination Options	N/A	Purchased through Known Payments	Renewal, Purchase at FMV, Return	Renewal, Purchase at FMV, Removal

services will be performed in what circumstances. While the price of the O&M may not be specifically broken out in some financing methods (such as PPAs), that cost is imbedded in the financing. Organizations that are willing to accept some responsibility (such as even being willing to have a local employee reset an inverter, if needed) may be able to reduce the costs.

General Contract Terms

PPA and lease contracts can have many clauses that are cause for concern for the customer, such as force majeure and removal costs. It is critical that all aspects and responsibilities of the contracts are clearly understood and agreed upon.

Rebate Applications

While it is not imperative that a customer apply for a rebate before selecting a vendor, it is important to apply for a rebate as soon as the organization can commit to implementing a solar project. Since incentive funds are usually limited and often tiered (and are often quickly oversubscribed), it is important to file a rebate as soon as a decision is made to go solar. Qualified integrators should be able to manage the entire rebate application process by providing the necessary forms and calculations, and even submitting on behalf of the customer. Therefore, a rebate application—especially those that require a deposit or fee—is a clear indication internally (and to integrators) that the organization is serious about implementing solar PV.

STEP FOUR: ENGINEERING & PERMITTING

Once the design, financing, and contracts have been finalized, the real work of the project can really begin. Engineering the installation advances the basic design by specifying all the details, large and small, of the hardware. At the end of this stage, one should see a finished system design with detailed design specifications and final engineering drawings. Plans finalized at this stage are the ones submitted for final permitting to the local building authorities. While this process can be fast-tracked, a typical project that is professionally managed and that engages and solicits feedback from the customer should take 1—3 months. Permitting issues can prolong this stage.

STEP FIVE: PROCUREMENT

With so many integrators offering solar services, and so many manufacturers of components offering various types of hardware, it is important for any customer to understand the basics of what works, why it is appropriate for their site while some similar hardware is not, and where to get it for a fair price. One will need to rely on the integrator for the details (almost all integrators will source the components, removing this burden from the client). Close, honest communications between the integrator and the client are essential here, because the performance of the system and the successful execution of its financial projections depend on choosing the technology that precisely suits the energy-producing potential of the site.

Some integrators might have a vested interest in selecting certain technologies over others, such as in the cases of integrators with long-term supply agreements or vertically integrated manufacturers and integrators. Full disclosure here of why technology selections are being made is important. Ultimately, the best choice is made based on the client's needs, not on the integrator's preference. Know why your integrator is making certain sourcing decisions over others. Typically, integrators will specify the products from manufacturers who provide them the best pricing. Since most products in a technology category are fairly similar in terms of performance and warranties, minor differences in efficiency, appearance, and price are the only major reasons to request a specific brand. In today's supply-constrained market, larger integrators or those that have longer track records are able to secure better pricing.

The design, engineering, permitting, and construction processes are very dependent upon the exact components specified. Therefore, there sometimes is an iterative process between the client and integrator whereby multiple scenarios are investigated prior to final design or engineering. Minor differences in size or efficiency in components can cause major redesigns and/or have noticeable impacts on project economics. Since the major components (the PV modules, inverters, and mounting system) can take from 0 to 4 months to receive from the manufacturer (even for vertically integrated manufacturer/integration companies), it is necessary to start this early.

STEP SIX: CONSTRUCTION, COMMISSIONING, & REBATE REQUEST

If the prior stages have been completed successfully, installation of a solar system should be a relatively standard process using standard construction techniques. By the time an integrator actually breaks ground, delays, interruptions, supply issues, permitting, or other problems should have been anticipated and overcome.

A typical 100—1,000 kWp project should take 6-12 months to complete from contract signature to final commissioning. Factors that impact this schedule (other than anticipated or unexpected delays) are contract requirements, site conditions, system size, local permitting requirements and utility interconnection procedures. The actual onsite construction time required to install a fully engineered system (not including closeout/punch list items) typically takes 4-16 weeks.

Actual construction includes: PV installation—including mounting system and module placement, bonding, and wiring; electrical installation—including inverters, DC disconnect, and piping; and then system interconnection and testing. The final installation step is commissioning, which includes complete system testing, final inspection, and interconnection with the utility grid. Once the system is complete to the integrator and customer's satisfaction and has passed all the integrator's tests, the building, utility, and rebate program administrator inspectors must verify the installation. After any required modifications have been made and verified, the paperwork for any incentives can be submitted.

STEP SEVEN: OPERATION AND MAINTENANCE

The most important part of O&M is understanding exactly what preventative measures and repairs may need to be done and who will responsible for each of them. If the system is financed with a PPA or if the customer purchases a good O&M service plan, then the operation and maintenance of a well-designed solar PV system should be the simplest part of this process—nothing. Most states require integrators to offer at least a basic break-fix warranty, but all qualified integrators should offer some type of service plan(s).

If the customer is responsible for O&M, then it is important to get some training from the integrator. The integrator should provide an O&M

package that clearly lists what maintenance should be performed when and how to do it. There are no moving parts, so simple visual inspections are the basic key to problem-free operations. Maintenance can be as simple as periodic system inspection to look for damaged modules, corrosion or loose electrical connections. Nearby brush or trees should be trimmed as needed to prevent shading. By now the customer should be well-acquainted with the warrantees of all components (whether from suppliers, dealers, builders or manufacturers), so that in the unlikely event of problems, the customer should know exactly to whom they should report.

Monitoring and data reporting are a key part of the O&M process. Many incentive programs and all PPA financiers will require an internet based monitoring system that tracks system production and ambient weather. In addition to alerting the system operator to underperformance, monitoring systems report system production that is used by PPA financiers and incentive program administrators to calculate the appropriate monthly bills or credits—similar to a standard utility meter.

Optimizing Performance

Regardless of whether the owner is the customer seeking to maximize energy savings or a PPA financier seeking to maximize rents, the project economics (and often incentives) are dependent upon optimizing the system performance at a reasonable cost. The metric of installation dollars per kilowatt hour estimated in year one is the single best measure of value optimization. Some integrators will offer performance guarantees to the system owner (i.e. under a purchase or lease, but usually not to the host customer under a PPA, since the host has no production risk). Performance guarantees usually come at a premium, so they should be evaluated and decided upon according to the organization's preference for warranties, its view of the risks, and its ability to assume some basic system O&M responsibilities. The most common system failure is due to a tripped inverter, which can easily be reset in seconds by an individual onsite.

While many contend that modules must be rinsed or washed regularly, that is often not economically sound. Regular rains are often sufficient most of the year. In dry or dusty areas, or in cases or times where the PBI is high (such as during the term of a performance based incentive), it can make more sense to wash the modules. Cleaning a large system typically costs $1,000 - $10,000 depending upon the location, availability of water, size, and system type.

VENDOR SELECTION

Once the project champion and decision makers understand the PV project process and options, it is imperative that the project goals and vendor selection process are developed and clearly communicated internally and to the integrator(s). Potential goals include: design aesthetics—such as visibility; dual functionality—such as carports; maximizing energy or carbon offset; and, of course, optimal project economics (IRR, ROI, NPV, or Payback). Most customers want a combination of all of the above, and it will take an experienced integrator or consultant to outline what the various design options will offer and cost in order to aid in goal prioritization.

Identifying an integrator (or a short list) can often be as simple as asking for recommendations or looking online for the state solar industry association. The solar integration process is a design-build construction project—in that there is a design/architecture, engineering, financing, and construction phase—and so it is important for the customer to engage the final integrator as early as possible. Since not all integrators can offer all financing or technology/design types, the earlier that the customer can identify the financing and design options that are most likely to fit with their facility and organization before selecting the integrator(s), the better. However, since most PV system customers are not solar technology or finance experts, it is common for organizations to do one of the following:

- *Sole Source to a Recommended Integrator*—This is the simplest scenario, and obviously is preferred by integrators (reduced cost of sales and competition leads to higher margin). In the event that a customer has a definite need for a particular proprietary product and/or is comfortable with the pricing arrangement provided (such as a cost-plus contract), then this is often the fastest method.

- *Bring in a Consultant*—A solar consultant can manage the process of initial assessment and identifying the best integrator(s), but at a cost.

- *Conduct an RFI or RFP*—While conducting a request for qualification (RFQ), or request for information (RFI), or request for proposal (RFP) can be a time consuming process, it is a simple way to solicit qualifications or bids from multiple vendors; this will educate the customer to the range of costs and design options, and increase the price competitiveness.

Every project must begin with a champion, and solar is no different. The solar project champion will have to manage the project from vendor selection, design, and financing through to construction, final rebate submittal, and O&M implementation.

When finalizing integrator selection, one should conclude the research by interviewing the candidates, and weighing their respective reputations, references, warranties, designs, effective $/kWh cost (which is more relevant than $/W), and projected energy output of their designs. One can use NREL's PVWatts or another solar production calculator to compare system performance, though these models are often complex and do not work for all technologies. If in doubt, check the status of their license(s) with your state contractor licensing board. Since integrator's components, designs, modeling methods, and financial and technical assumptions will vary, one should make an effort to understand the variations and effects or request highly specified proposals in order to be able to be able to compare between integrators.

TRENDS AND INNOVATIONS

While there have been many innovations in the last decade, the exponential growth of the solar PV market is driving innovations in all aspects of the industry. The trends range from more comprehensive environmental policies, to greater system design/technology offerings, to creative financing methods. However, while some people suggest waiting on the sidelines until a technology breakthrough happens, that may not be necessary or prudent for the following reasons:

1.   PV has been developed for over five decades and through many technological avenues—the bottom line is that commercially viable PV products take a lot of R&D to come to market, and even then, many of the new technologies are only marginally more cost effective than their predecessors.

2.   The incentive programs are designed to decrease as costs decrease with the end result being that a PV purchase today and five years from now should have similar economic benefits for a customer. However, incentives often drop quicker than prices as rebate programs become oversubscribed, and PV module costs have actually gone up from

2004 through 2008 (due mostly to high demand in Europe).

3.   Manufacturing and installation cost reductions may not be realized due to global market dynamics and supply economics (why sell it cheaper in one market, when one can get more in another).

4.   The public relations value (such as press coverage) of PV projects will decrease as demand and regulation increase.

Some of the most relevant trends include the following:

Market Building Incentives

Put simply, right now there are almost no economical projects without incentives—period. Germany made a commitment to become energy independent and they now enjoy the largest renewables industry in the world. This was through the use of a national feed-in tariff, which is what many European countries have elected to use and what we may see in more use in the US soon. More and more uniform federal and/or state incentives—such as a long-term extension of the federal investment tax credit—will create green collar jobs, reduce the cost of renewable energy, and increase our energy independence.

Financial Innovations and the PPA

The advent of the third-party ownership finance (no up-front capital cost) Power Purchase Agreement in 2005 (made possible by the 30% ITC in the Energy Act of 2005) revolutionized the industry and is responsible for spurring much of the recent growth in the PV industry.

Solar PV Installation Costs

Increased volume throughout the value chain, coupled with hyper competition and vertical integration will continue to bring installation costs down. Typical prices today range from $6-10/Wp.

New PV Technologies

There are many exciting technologies in the works, such as Concentrating PV (CPV, where less expensive lenses or mirrors are used to reflect more light onto less PV material, such as is being pioneered by companies like Energy Innovations), thin film (which uses less of the expensive semiconductor materials with manufacturing processes that can be less expensive, such as Uni-Solar, First Solar, Applied Materials, and many others).

While they usually take longer than anticipated to come to market, the more successful technologies and companies will eventually enter the market and aid in overall PV cost reduction.

PV Efficiency Improvements
Currently available crystalline modules are 12-20% efficient and the theoretical efficiency limit (for a single band gap, or layer, of PV) is about 31% (by layering multiple junctions of photovoltaic material with specific light spectrum sensitivities, manufacturers have been able to achieve upwards of 40%). While a doubling of power output will help project economics, the associated cost of that R&D will be felt in the module prices of higher efficiency cells.

Tracking Systems
In addition to higher efficiency and lower cost systems, another way manufacturers are designing more productive systems is by developing single and dual axis tracking systems that can be mounted on roofs and carports/canopies, thereby greatly increasing the system production factor.

Installation Component & Method Improvements
The mounting and racking systems for PV are being improved to enable faster installation, with fewer parts, in more standard configurations—all of which should help to lower the install cost.

Grid Parity
Between rising utility costs and reduction in installed prices due to scale, the PV industry is on track to reach grid parity at some point in the next decade (depending upon insolation/incentives/utility rates). Organizations that go solar sooner will enjoy more savings (and more PR) and other benefits sooner.

Cleantech Investment
Investment in clean technologies—such as alternative energy and other sustainable manufacturing and building processes—and the companies pioneering them are currently the largest investment growth sector. Many organization are realizing the benefits, savings, and opportunities associated with more sustainable practices.

Climate Change and Green House Gas (GHG) Regulation
As the world becomes increasingly concerned about global warming,

there will be increasingly stringent emissions control/reduction/elimination policies enacted; the result will be a requirement for more renewables and increased costs for fossil fuel generation.

Renewable Energy Credits and Carbon Offsets

RECs and carbon are similar in that they both attempt to provide a market-based approach to valuing the negative environmental externalities associated with fossil fuel based electricity generation. Expect to see REC and carbon markets continuing to develop as more legislation is developed and more demand arises from an increase in sustainability awareness.

Energy Efficiency

EE has always been an important industry, and although solar integrators are eager to sell a PV system, negawatts (demand reduction) are cheaper than megawatts (of PV). Expect to see increased EE penetration towards the commercial scale market as solar incentive programs increasingly require integrators to perform at least cursory EE audits. Packaged EE + PV project financing is a nascent but growing industry.

Solar Btu

Just as the market for solar electricity has recently been reinvigorated (since the last oil crisis in the 1970's), we should expect to see more solar thermal space, process, and hot water heating technologies and financial offerings. Like EE, this too can be packaged with PV.

Utility Rates

US average commercial rates went up over 9% from 8.67 cents in 2005 to 9.46 cents in 2006—the last year of released data, and the energy rates will continue to rise as international demand continues to skyrocket (driven by China and other areas undergoing high growth and industrial expansion) and environmental regulations increase compliance costs.

Utility Tariff Designs

Utility tariffs for commercial scale customers often have relatively high demand charges and, therefore, are not very solar friendly. As the industry grows, one inexpensive way to improve project economics is for Public Utility Commissions to require—or for utilities to choose to offer—"solar friendly tariffs" that can potentially increase the average avoided cost of power offset by solar, thereby increasing the savings from the same system (this is more true or apparent for PV systems that offset a higher percentage of usage).

SOLAR PV SYSTEM WORKSHEET

The following will provide rough estimates of PV system size and area needed to achieve your offset goals.

Estimate PV System Production Factor (kWh/kWp/yr)
1.   Pick the appropriate solar radiation value from one of the NREL charts (either fixed tilt or tracking, range: 4-9): _____ kWh/m²/day

2.   Determine the Production Factor (range: 900-2025):
225 x (solar radiation from above) = _____ kWh/kWp/yr

Estimate PV System Size Needed (kWp)
1.   Annual kWh usage (from your utility bills): _____kWh

2.   Chose Your % Energy Offset Goal (max ~ 90%): _____ %

3.   Determine kWh Offset Goal: (% Offset) x (Annual kWh) = _____kWh
4.   Determine System Size Goal: (kWh Offset Goal)/ (Production Factor) = _____kWp

Estimate Unshaded Space Required (sqft)
1.   Either assume 10 Wp/sqft Power Density (and skip to step 3)—or—Pick a tilt between 0-30 degrees: _____

2.   Determine Power Density: 13 Wp/sqft—(Degrees/4) = Wp/sqft _____

3.   Determine Area Required for a Crystalline Non-tracking PV System: (Estimated System Size kWp) x (1000 W per kW)/(Power Density) = _____sqft

4.   If Thin Film or Tracking, Double the Area: 2 x (Above Area) = _____sqft

What You Need to Provide Your Integrator (checklist)
✓   12 months of electricity bills (using a utility bill release form or faxed/scanned copies)
✓   Site map with a written or marked description of meter location(s) and possible areas that can be allocated to a solar PV system for 10-20+ years
✓   Roof as-built drawings with elevations, types, ages, and conditions (for roof projects)
✓   Basic information regarding location(s), ownership/organization type, credit, and financing preferences

Factors to Consider When Evaluating a Solar PV Proposal
✓   Integrator's overall experience and length of time in business
✓   Integrator's specific experience with the type of project proposed (design and technology)
✓   Positive reference checks
✓   Team—the project will only be as good as the actual people working on it
✓   Equipment and warranty—is it custom, a common industry brand, or a proprietary product
✓   California Energy Commission listed modules & inverters (most incentive programs require this)
✓   Design—optimized for your technical, financial, aesthetic, and location requirements
✓   Clearly listed, defined, and explained technical and financial assumptions
✓   Price is important, but change orders, delays, and failed financings are common for lower bidders

SUMMARY

There are many compelling reasons for implementing a commercial scale (10—2,000 kW) solar PV project at your organization's facility now, or in the near future. Many companies, government agencies, and non-profit organizations are finding that the current combination of environmental, societal, and economic benefits provided by solar PV projects make it easy to choose to go solar. The recent surge in incentive programs and innovation in technical and financing options—such as thin film and the no capital upfront solar Power Purchase Agreement model—are providing customers with a variety of options that simplify and facilitate the solar PV project process. The steps described above and the supporting information should prepare the reader to be their organization's solar champion and to implement a highly successful solar photovoltaic power project.

References

Database of State Incentives for Renewable Energy (DSIRE) www.dsireusa.org
Solar Energy Industries Association (SEIA)
SEIA Guide to Federal Tax Incentives www.seia.org
Solar Electric Power Association (SEPA) www.solarelectricpower.org
California's Consumer Energy Center (CEC) www.consumerenergycenter.org/renewables/solar/index.html
CEC Eligible Modules & Equipment www.gosolarcalifornia.org/equipment/index.html
National Renewable Energy Laboratory (NREL) www.nrel.gov/solar
NREL PVWatts Solar Calculator www.nrel.gov/rredc/pvwatts/
NREL Solar Resource Maps www.nrel.gov/gis/solar.html
U.S. Department of Energy (DOE) Photovoltaics Program www1.eere.energy.gov/solar/photovoltaics.html
U.S. DOE Energy Information Administration www.eia.doe.gov
U.S. DOE Electric Power Annual Report www.eia.doe.gov/cneaf/electricity/epa/epa_sum.html

Chapter 5

Green Transportation

Arjun Sarkar
Transportation Department
University of California—Santa Barbara

INTRODUCTION

Around the country today in boardrooms, over lunch and on golf courses, sustainability has become a new buzzword, influencing the conversation regarding the long-range viability of given business plans, products and services. A new pioneering spirit hearkens back to that which infused the country's founders, with a new "can do" attitude regarding energy and sustainability.

Why are they doing this? Because, right now, our transportation is a mono fuel (oil) dependent system. 97% of the transportation sector relies on oil as its energy source. This system taxes us in many ways: energy security, foreign policy, local pollution and global warming. With strong impacts on the economically that require us to rely on imported oil to keep the U.S. moving.

Whatever product, service or business plan we might be working on we depend on the current transportation system. We are seeing the volatility in price as well as the vulnerability of supply today. Below are a few stories from locations in California that have taken on the challenge to transition to "advanced fuels" (not "alternative fuels") in the transportation sector.

INSPIRING SUCCESS STORIES

Santa Barbara, California:

The Metropolitan Transit District has been running battery electric buses since the summer of 1991. The choice to take on this project was controversial at the time of inception, due to the technical challenges. At the time they were running diesel buses down the main street "State Street," a "shopper hopper" service. Between the exhaust and the noise, something had to be done.

Battery electric buses were not being manufactured at that point. That meant the RFPs that went out were going to companies that were start-ups, and they were "forcing" the technology. This commitment, led by the general manager, Gerry Gleason, was a major undertaking. It took one or two people with vision and leadership to bring the concept to fruition.

The electric program is still going strong in Santa Barbara today. It has accumulated over 2.8 million miles on pure battery electric, and there are 22 electric buses in the program. The diesel fumes are a memory from the past, and the city's residents take strong pride in their city's fleet.

Palm Springs, California:

Sunline Transit Agency parked their diesel bus fleet one night in 1994, and in the morning the drivers drove out in a new compressed natural gas (CNG) bus fleet. They were the first agency in the US to stage such a dramatic transition to an advanced fuel. The man with the mission was Richard Cromwell (general manager at the time). Richard Cromwell then started working on a hydrogen fueling station. By 2000 Sunline had a hydrogen fueling station on the premises that was being powered by a photovoltaic array.

Sunline was a founding partner in the International Fuel Cell Consortium. That year Sunline also became a California Fuel Cell Associate Partner. In November 2006, Sunline opened the first large-scale hydrogen facility in the United States, using commercialized technology available for third-party refueling purposes. The city of Palm Springs (and surrounding cities and agencies) switched their fleets to CNG, and the local College of the Desert has a program dedicated to alternative fuels. Again, these major transitions came to pass because of a vision and mission undertaken by one district, and championed by one or two people.

Fresno, California:

Fresno is in the San Joaquin Valley, one of the two worst polluted air districts in California. Los Angeles and the San Joaquin Valley vie for this title. For years they tried to put together an alternative fuel program in the area. It just kept falling apart.

Then came Allen Autry, as a pro-football player, actor-turned-politician, and now mayor of Fresno. He brought the players and the community together and created an "Alternative Fuel Playbook," with spectacular results. The city of Fresno was selected for a national award at the 11th Annual Department of Energy Clean Cities Conference, over 88 Clean Cities Coalitions across the nation (comprised of 4,800 public and private sector agencies), in recognition of its leadership in the implementation of an innovative "clean air" fleet program. Mayor Autry accepted the award at the conference, held in Palm Springs on May 3, 2007, with over 1,000 stakeholders in attendance.

Fresno was recognized for its use of varied alternative fuel technologies within its fleet, such as liquid natural gas (LNG) refuse trucks, CNG transit buses, exhaust after-treatment devices on heavy equipment, and gasoline-electric hybrid vehicles. Through the use of these technologies, the city of Fresno displaced over 800,000 gallons of petroleum use in 2004.

As you can see from these stories, the key factor in creating high-impact results lies in a few individuals stepping out and taking initiative, creating projects and pursuing partnerships that ultimately cause a shift that alters the future of transportation.

> *"Never doubt that a small group of committed people*
> *can change the world; indeed, it is the only thing*
> *that ever has." — Margaret Mead*

ADVANCED FUEL CHOICES

Before choosing a fuel, ask these questions:

1. What is driving you to consider an advanced fuel? Is it philosophy? What are your goals? Examples: domestic energy, clean air, or reduction of CO_2.

2. What is the vehicle going to be used for specifically? Include in this consideration the load, distance traveled, and the overall route.

3. What is the task at hand? As with computer technology, don't buy more than you need.

4. How cutting edge do you want to be?

5.   Do you have too many vehicles in your fleet now? Assess what the
      vehicles are being used for currently.

6.   Have you considered an incentive package for your employees for
      using public transit, carpools, and advanced fuel or Hybrid cars?

Biodiesel

In the U.S., biodiesel is mostly derived from soybeans. In other parts
of the world they are using palm oil, rapeseed and jatropha. It can also be
derived from animal fats. This fuel is registered with the Environmental Protection Agency it must meet ASTM D6751. The chemical name for biodiesel
is methyl esters. Biodiesel is not the same thing as vegetable oil. There is
a lot of confusion regarding this. I think it is because some people install
"kits" that allow them to run vegetable oil. This became popular at around
the same time biodiesel came to market. Vegetable oil is not a legal motor
fuel and is not recognized as such. The mantra that I recommend regarding
biodiesel is quality, quality, quality. Make sure that the fuel meets ASTM
D6751 and that the marketer as well as the producer is BQ-9000 certified.

There are many reasons to consider using biodiesel for your fleet. It
is a strait replacement fuel for diesel thus the ease of transition is a plus.
The reductions of emissions that are achieved from B100 are 67% reduction of unburned hydrocarbon, 48% reduction of carbon monoxide, 47%
reduction of particulate matter. Now the numbers for NO_x are being studied but there can be and increase in NO_x of up to 10%. There are post-com-
bustion treatment technologies that are coming to market that can reduce
the NO_x. When we look at the energy it takes to farm and transport we
find that for every unit of energy invested, there are 3.5 units of energy
gained. Biodiesel is creating domestic jobs and lowering our dependency
on foreign oil. You can use blends from 2% to 100% this is designated
by B2 being 2% biodiesel 80% diesel and B100 being 100% biodiesel. In
terms of warranty with the individual engine manufactures you need to
check into what they will warranty. One good location for information is
*The National Biodiesel Board (NBB) www.biodiesel.org on their web site
they have information on some of* the engine manufactures. The one thing
here is the local dealers most likely will need to be educated on what the
corporate offices' official stance is on the warranties.

E-85 Ethanol

The main feedstock for E-85 here in the U.S. is corn; in other countries they are using sugar cane as well as sugar beets. There are a lot of conversations going on currently regarding ethanol. I will give you some information that will help in this dialog. At this time there are companies that are using sugar waste from the candy industry as well as the spirits and soft drink industries. These waste sugars would either go into the sewer or landfill. It is a small amount at this point around 5 million gallons of ethanol a year but it is a concept that is working today. Another project that I find interesting is a company positioned their new ethanol plant in the center of CA cattle industry Madera CA. We in California import major amounts of corn as feed. The idea here is to squeeze the same corn that is going to the cattle for ethanol and send the pulp/carbohydrates on to the cattle. This in turn is good for the livestock as not to have the high sugar content. Here we have a fuel that the manufactures have built cars specifically to run the fuel. The designation of those vehicles is "flex fuel." They can run regular gasoline 100% to 15% gasoline and 85% ethanol E-85.

A vehicle purchased today will be with us for 12 to 15 years in the used market. A flex fueled vehicle that is running corn E-85 today in 5 to 10 years the E-85 may be cellolosic ethanol or from nano-technology/genetically engineered forms of algae. Studies and critical thinking are important to make these choices but as end users we need to get the job done. The US economy does not stop, goods and services must keep moving. These fuels are in the marketplace now and the vehicle manufactures have the vehicles available. It is our job to chose the fuel type create a program and implement it.

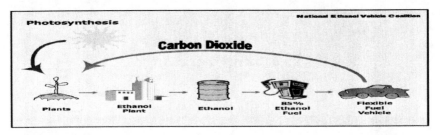

Natural Gas

Mostly consists of methane but also contains some ethane, propane and other gases such as nitrogen, helium, carbon dioxide, hydrogen sulfide, and water vapor. Natural gas is a great motor fuel, it is called compressed natural gas or liquid natural gas CNG & LNG respectively. CNG/LNG has a high octane of 130. On board a vehicle CNG is stored at 3,600

PSI, LNG is stored in a liquid form produced when natural gas is cooled to minus 259 degrees Fahrenheit through a process known as liquefaction. Natural gas is delivered to home and commercial buildings throughout the US. For CNG the only thing that needs to be done with the natural gas is removing the water and compressing it. There are incentives at the Department of Energy level as well many states have incentives that can be combined with the federal money. At this time CNG makes economic sense environmentally and I would say equitable as well, as it lowers our reliance on foreign oil. One thing that is often overlooked when talking about natural gas is the fact that it's methane and that methane can be produced from biological sources. Sweden has about 13,500 CNG vehicles, and more than half of the CNG is bio-methane. We can obtain our natural gas from many sources, like wastewater treatment plants, the cattle industry, biomaterial, landfills, and others. As we use more of these advanced fuels and look to renewable sources, our energy security increases.

Below is a picture of a "bio reactor," which generates methane from manure.

Hybrids and Hybrid Plug-in

Hybrids have been in the marketplace for some time now. The first year that a production line car was sold was in 1999 the Honda Insight. It got an EPA rating of 60/66 mpg. Hybrids have come a long way in a

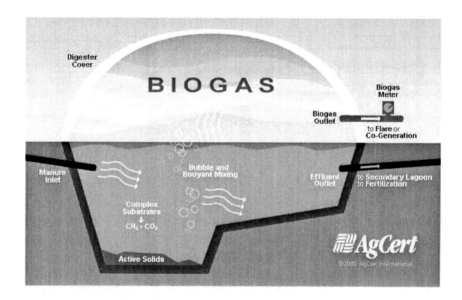

short time. The basic theory of a hybrid is that it uses two power sources. Toyota and Honda have had the lion's share of the market for quite some time now. Ford's hybrid Escape was the first hybrid SUV. GM, Chrysler and BMW all worked together to create a dual mode hybrid transmission. This unit has a low- and high-speed electric continuously variable transmission (ECVT) modes; the system is commonly referred to as the 2-mode hybrid. Hybrids are increasing our MPG numbers and they are keeping the technology moving forward.

When you add the advanced fuel component to the hybrid now you have something exciting. Say an E-85 hybrid or a hybrid diesel that we can run biodiesel in. We are seeing these as concept cars and I am sure that they will come to market at some point. Regarding this concept there is one getting a lot of attention at this time—the **plug-in hybrid**. In this case there is a larger battery pack than in a regular hybrid. The range that most people are targeting is around the 35 to 40 mile range on a single charge. If you combined this with time of use electric meeting it starts to look very good. The Chevy Volt has been getting the attention on this front. In their concept the system is a "flex" fuel a little different take. You could order it as E85, Biodiesel or have it be a fuel cell plug-in. Ford is working with Southern California Edison looking at how the electric grid and plug-in hybrids can work in a synergistic system. At this point in time no major manufacture has a plug-in car for sale. There are companies that will sell conversion systems for existing hybrids.

Electric Vehicles EVs

Battery-powered electric vehicles are designated as zero emission vehicles (ZEVs) because they have no tail pipe emissions. However, there are upstream emissions associated with the electricity. Thus the grid make up of the state that you are in is going to depict the CO_2 and the renewable content of the electricity. The electric grid today has almost no oil in the mix. Battery electric or plug-in hybrids can use the grid today. The diversified portfolio of the grid makes this choice very appealing, with the onset of new smart home electric meters that bill you at varying rates though the day. If you charge your vehicle at night when rates and usage are low this can be a win-win. Here are a few key points regarding EVs. If you have renewable electricity at your site, let's say from Photovoltaic (PV). The owner of said property has over-specked a system or has done energy efficiencies to the home after the system was installed. A second or third vehicle that is an EV can utilize the renewable electricity. This off the shelf system starts to get to the sustainable transportation transition. The efficiencies of EVs with motors being more than 90% efficient and batteries at 75-85% operating costs can be reduced. The inherent nature of an EV drive train with so few moving parts can lower service costs. Without the engine oil or coolant the fluids that can contaminate soil and water (multimedia impacts) are limited. The main two categories of EVs are full function; these can travel at freeway speeds and can take the place of a regular vehicle. The other is a neighborhood electric vehicle (NEV) and this vehicle is limited to 25MPH. There is a third type that is called a city car but at this time no manufacture is making one. Also there is no official vehicle code for a city car.

Hydrogen H_2

Hydrogen is the simplest and most abundant element in the universe. This gas has great potential in the coming energy transition. It is a true energy storage device that can play a vital role in the ebb and flow inherent with various renewable energy sources. For all of these energy carriers biodiesel, ethanol, natural gas and electricity. It is important that we see these as end fuels or energy carriers. If the upstream inputs/raw material is diverse then the fuel is not tied to one commodity. I will elaborate on this a little more. The electric grid has many types of energy inputs coal, natural gas, nuclear, geothermal, hydro, wind and solar. With ethanol the feedstocks could be sugar cane, sugar beets, wheat or waste material. If one commodity spikes in price or there is a supply problem you are not relying on just one feedstock, as is the case with oil. Hydrogen is probably the most diverse fuel when you look at the upstream possibilities of what you can produce hydrogen from. I have attended many conferences, classes and seminars on hydrogen. The question of pathways to generate hydrogen is not the main hurtle to the hydrogen economy. The things that I see as barriers at this point are hydrogen storage, fuel cell stack durability and infrastructure. Regarding storage the new Honda FCX Clarity has a range of 270 miles at 5,000 PSI. Also looking at infrastructure Honda has proposed a home refueling station as they have for CNG (Phill). GM is using a 10,000PSI hydrogen system on their vehicles to help with the storage issue. BMW is going a different direction with a liquid hydrogen storage system and running it not through a fuel cell but rather an internal combustion engine. I feel that we are at the point that we were at back in 1999-2000 with the onset of the hybrid vehicle. The wave is coming and a few manufactures are going to be ridding it. There are opportunities to demon straight this technology, are you a "techie" company? Do you want to play?

For More Information
This website is a great resource-Drive Clean California http://www.driveclean.ca.gov/en/gv/home/index.asp

California is implementing a new system for labeling vehicles. The new smog index label will include a global warming component. It will be easy to use a 1 to 10 scale with 10 being cleanest. CA is in litigation with the U.S. EPA regarding its ability to regulate CO_2 emissions from vehicles. Regardless of this suit automakers will be required to meet the new labeling rules. Eleven other states, including New York and all three west coast states, are prepared to adopt California's new regulations, so the labels would likely be included on vehicles for sale in those states as well.

DOE Incentives for Advanced Fuel
    http://www.eere.energy.gov/afdc/incentives_laws.html

DOE fleet portal, which replaces the familiar "Clean Fleet Guide," is a one-stop-shop for fleets
    http://www.eere.energy.gov/afdc/fleets/index.html

Energy Independence Now!
    http://www.energyindependencenow.org/

City of Santa Monica
    http://www.smgov.net/epd/scpr/Transportation/T7_AlternativeVehicles.htm

Securing America's Energy Future
    http://www.secureenergy.org/site/page.php?index

Recommendations to the Nation on Reducing U.S. Oil Dependence
    http://www.secureenergy.org/files/files/147_ESLC_Oil_Report.pdf

CALL TO ACTION

We are seeing the pressers of living with a mono fuel paradigm every day in the media. We are setting record high oil prices at over $100 per barrel. There are hidden and indirect costs that are almost incalculable.

The pioneering spirit that founded this great country is still here and alive in each and every one of us. The bottom line is but one measure of our success. During the short time that we are here, to leave our mark, what are we going to do with that precious gift? I urge you if you are a fleet operator, energy manager or CEO develop an advanced fuel plan and begin to create a transportation fuel portfolio. Thank you for all that you do.

Chapter 6

Water Efficiency Measures

Bill Jacoby
Bill Jacoby Water Resources Consulting
billjjacoby@aol.com

BACKGROUND

For a variety of reasons it is becoming increasingly important for commercial and industrial water users to consider ways to increase the efficiency of their water use. This chapter will explore the reasons why smart managers are increasing water use efficiency, the types of water use with a high potential for efficiencies, and specific measures proven to increase efficiency.

While the Southwestern and Western United States have for years been concerned about water availability, other parts of the country are more frequently faced with those same issues. Climate change is a much-discussed issue in the world today and this chapter makes no attempt to contribute to that discussion. However, prudent utility managers are considering the impact recent trends in weather patterns have had on water supply and the availability of hydro generated power. With this uncertainty in mind, it makes sense to consider what steps your business can take to increase water use efficiency. This is especially true of industries that require a reliable supply of high quality water for production purposes. For example, during the early 1990s when the San Diego region was faced with a 50% reduction in water supply, the impact was huge for the manufacturing industry. The ability of firms to manufacture everything from computer chips to ingredients for beer and foods to lifesaving medical products was at risk. Fortunately, late winter rains and intense water conservation efforts helped avert industry shutdowns.

The negative impact an unreliable water supply can have on a community, state and even the nation has become increasingly apparent. Accordingly, water agencies have teamed with federal and state government to take steps that will increase water use efficiency. Some of those efforts have been through standards for water efficiency in new toilets, faucets, urinals, and showerheads. More recently, some states have adopted stricter standards for new commercial clothes washers and dishwashing equipment. Another approach is to encourage the replacement of old inefficient fixtures and appliances through financial incentives. The incentives are typically funded by agencies that will benefit by the decreased use of water (i.e. water and sewer agencies, energy suppliers, and state as well as federal government agencies.) Specifics examples of these incentives will be provided later in the chapter.

Water has usually not been one of the major cost considerations for many firms, while the cost of energy has been more often of concern. However, we are seeing changes in water pricing that are increasingly getting the attention of commercial and industrial water users. First, most water agencies are moving to pricing structures that include a commodity charge as well as fixed charges. Some are also considering inclining rate commodity charges. In other words, the per gallons cost of water will increase as consumption increases. Water agencies are developing "water budgets" for landscape water use that can be used as the basis for determining the inclining thresholds. Consumers also know that in many cases saving water reduces energy demand. This is especially true where hot water is involved. Finally, it is becoming increasingly clear that the power needed to move and treat water is a major consumer of the limited capacity of the country's energy producing plants. In California those uses account for 19% of the state's energy use. The increased competition for limited power, especially during peak demand times, will tend to cause power shortages and eventually drive up the cost of energy (and water) for all.

Sewage disposal costs also need to be calculated when considering the cost effectiveness of water conservation measures. In some communities the cost to businesses for sewage disposal exceeds the cost of water purchased. The sewage bill can be based on both the volume (usually determined by water consumption) and level of contamination of the effluent discharged. For firms that use water in a process resulting in a heavy contamination load for the discharged effluent, the sewage cost is often substantial. Water use efficiency can certainly contribute to deceasing the volume of effluent, and some process changes can save water and reduce contamination load levels.

Finally, water use efficiency is a way for firms to demonstrate a commitment to enhancing the environmental quality of their community. Water not used by businesses can stay in our lakes, streams, and groundwater supplies. Additionally, by reducing sewage flows we decrease the amount of treated sewage returned to those lakes and streams. In fact, some businesses may want to showcase their commitment to water use efficiency. These demonstrations of conservation measures already taken may prove invaluable during times of water shortage when water agencies are determining which customers will be called on to absorb water delivery shortfalls. Finally, water use efficiency can be simply a good public relations asset for your firm that will pay off in demonstrating that you are a good neighbor in your community.

WATER CONSERVATION MEASURES

The 20 water conservation measures identified in this chapter have been broken down into four types of uses.

- **Interior water use** is defined for purposes of this chapter as use in plumbing fixtures located in existing business structures.

- **Landscape use** includes all water used for landscape irrigation and the measures to increase efficiency are presented in two categories: landscape design, and landscape installation and maintenance.

- **Pools and fountains** include swimming pools, spas, decorative pools, and fountains.

- **Process** use includes any water used in facility cooling or for manufacturing processes.

Interior for New Construction and Retrofits
1. Prior to the early 1980s there were not any water use efficiency standards for toilets and showerheads. That changed in the early 1980s when the US government adopted requirements that toilets either manufactured in the US, or imported, not use more than 3.5 gallons per flush (GPF) and showerheads not more than 2.75 gallons per minute (GPM.) These requirements were strengthened in the early 1990s when the per flush standard was lowered to 1.6 GPF leading to the ultra low flush toilet (ULFT.) The showerhead standard meanwhile was reduced to 2.5 GMP. Technological advances have

produced products that exceed these standards and many are gaining popularity.

Because of these efficiency standards, toilets installed since about 1992 are likely to be ULFTs and showerheads installed since then can also be assumed to be fairly efficient. However, as noted above, water supply agencies in some parts to the nation are offering financial incentives to encourage businesses to retrofit pre-1992 fixtures with more efficient toilets and showerheads. For instance in Southern California businesses are offered up to $135 for retrofitting with a ULFT and $165 for a toilet that uses fewer GPF than the 1.6 GPF standard. Free or low cost showerheads are sometimes available through either local water or energy suppliers.

2.   Urinals installed before 1989 use 3 or more GPF, compared to the current industry standard of .5 GPF. In fact, there are urinals now available that use no water. Incentives from water agencies can range up to $200 per water efficient to $400 for zero-water urinals.

3.   Faucet aerators have improved in efficiency at the same pace as showerheads. Simply put, they inject air into the water stream to produce a flow of water sufficient to wash hands while not using nearly as much water as a solid flow. Again, many local water and energy agencies offer these devices for no or little cost.

New or Replacement of Existing Landscape

Land use agencies are increasingly considering water efficiency requirements in design of landscape in new construction. As water shortages become more common, more consideration is also being given to replacing existing landscapes with more water efficient designs. In both cases, planning and design will require organizing the landscape for aesthetics, practicality, and water efficiency. There are five major components to developing a good water efficient landscape design.

4.   There can be a place for nearly any plant in a water efficient landscape design. It is important to use the concept of hydro zoning, or grouping plants according to their water and sun needs. There are many interesting and colorful low water use plants now available. Additionally, it may be appropriate to consider new concepts like artificial turf for high use areas like sports fields. Take advantage of all

these options whenever possible to create an inviting exterior for your facility that both visitors and employees will enjoy.

5.   Soil conditions can range from great black dirt to rocks and clay. A soil analysis should be performed to determine the type of soil where the landscape will be planted. The addition of organic matter to the soil increases water retention and penetration. Two to three inches of mulch will keep soil cool, reduce evapotranspiration and weed growth. Mulch should be spread on the surface around trees and shrubs and can be rocks, wood chips, bark, or shredded wood.

6.   Selecting the right irrigation system is one of the most important decisions in good landscape design. It will be nearly impossible to efficiently irrigate the landscape without the right system. In addition to saving water, an efficient irrigation system reduces irrigation water runoff and the pollution caused by the fertilizers applied to the landscape. Many recent innovations in irrigation technology enable slow, steady, and specific water application. Look for words such as "low gallonage" or "low application rate." Consider drip systems for shrubs and trees and always locate turf on a separate valve.

7.   Once the plant selection has been completed and zoned irrigation system designed, a water budget should be developed. The water budget considers variables like: soil type, plant water needs, irrigation system application rates, and projected seasonal evapotranspiration for each valve. Irrigation requirements identified in the water budget then can be translated into a seasonal, or monthly, schedule for each valve. The latest technology in irrigation controllers includes "smart controllers" that make the adjustments in irrigation schedules based on pre-programmed monthly evapotranspiration data by zip code.

8.   Alternative water sources such as reclaimed water, rainwater, or other options should also be considered. The potential for use of these options depends on local rainfall levels and availability of reclaimed water.

With a solid landscape design plan in place, proper installation and on-going maintenance are the next two key components to a water efficient landscape. Both of these require attention to detail and consistent monitoring to assure permanent efficiency.

9.   Selecting the right contractor, or staff, to install the landscape as designed is important. If the vegetation is not planted properly, or the irrigation system installed as specified, the property owner can't expect to achieve the water savings anticipated. Just as important is selecting the right landscape maintenance contractor. In both cases, it will pay to check references, look at pervious projects, and ask about training the contractor has in water use efficiency landscape installation or maintenance. Avoid the "mow, blow, and go" maintenance contractor and go with the firm committed to helping you reach your firm's efficiency goals.

10.   Before accepting the completed construction project from the contractor, make sure a comprehensive test of the irrigation system and inspection of vegetation installation has been completed. A good inspection will save you money and headaches in the long-run.

11.   The contractor retained to perform the on-going landscape maintenance must have a comprehensive knowledge of the vegetation, irrigation system, and plant irrigation requirements to be effective. It is wise to require that the contractor present a written plan specifically detailing how your landscape will be maintained.

12.   Consider using technology like waterbrooms to save water in hardscape maintenance. They can be used to clean tennis courts, pool decks, kitchens, outdoor eating areas, sidewalks, and other areas. Waterbrooms can save 5 – 15 gallons per minute over a hose and clean in about 75% less time. Additionally, food waste and other pollutants can be gathered and properly disposed of, rather than going into the stormdrain.

Pools and Fountains

13.   Fountains and decorative pools can provide a great addition to any landscape. In fact, as cultures in arid areas of the world learned long ago, these water features have a cooling effect. The sound of water is relaxing and creates an illusion of and oasis. Unfortunately, because these water features are so successful in creating that illusion of water plenty, they sometimes become an early target when water cutbacks are required. A good way to assure efficiency is to make sure the water used is recycled and used over and over. Better yet, if reclaimed wastewater is available in the area consider using that rather than

potable as the source to fill and replenish the water features. Attention to proper water quality care is also important as it can help reduce the need to drain and clean the feature as often.

14. Swimming pools and spas are great sources of recreation and provide a venue for exercise. Yet like decorative pools and fountains, they often become targets during periods of water shortage. Swimming pools should always be filled only to the proper level to avoid water washing over the sides. Again, attention to water quality will reduce the frequency of draining the pool for an acid-wash. Pool covers are an excellent way to avoid evaporation and reduce heating costs by keeping the water warm.

Process Use

15. Two opportunities to save potable water in the cooling towers used in building air conditioning have been identified. Cooling tower pH conductivity controllers lead to water and energy savings as well as longer equipment life. A properly functioning conductivity controller helps automated monitoring and control. This results in reduced scaling, bleed-off water and chemical use. An alternative is to use reclaimed wastewater when consistent with health department regulations.

16. Pre-rinsing dishes to remove large chunks of food prior to entering the dishwasher can use a significant amount of water. Until recently there were no water efficiency standards to pre rinse devices. By retrofitting with the new high-velocity, low-flow devices both significant water and energy savings can be realized.

17. National commercial clothes washer efficiency standards for new washers have focused only on energy savings. However, some states (including California) have set their own water use efficiency standards in addition to the national energy requirements. Additionally, managers of coin operated washers in: laundromats, apartments, college dorms, and condominiums have taken advantage of water agency financial incentives to replace older inefficient machines with new water and energy saving single and multi-load washers.

18. Medical X-ray processing machines had been notorious water wasters. The devices run water 24/7 even if processing was not taking place.

Each machine was using enough water in a year to meet the needs of 6 family homes. A simple recycling device is available to recycle that water resulting in significant water and sewage disposal savings.

19.   Commercial carwashes and on site vehicle wash facilities should recycle water from various stages of the washing process. For instance, water used in the final rinse should be saved and used in and earlier rinse, and ultimately as water in the wash cycle. If recycled water is available from a local purveyor it should be used. These steps will not only save on water bills, but significantly save on sewage disposal costs.

20.   A high quality reliable supply of water for its industrial process is obviously of great importance to any firm. While water and energy suppliers are often eager to advise these firms on ways to be more resource efficient, it must be remembered that modifying the industrial process to increase resource efficiency will be considered in light of many other factors. A few years ago I had the opportunity to meet with representatives to a major television manufacturing firm. To make certain it was clear where the firm stood one of the representatives started the meeting by saying, "How about if we don't tell you water agency folks how to supply our water and you don't tell us how to build TVs." This simple statement made clear to me that in the highly competitive world market, water use efficiency is not necessarily a major concern to manufacturing firms. However, sometimes there are ways to look at treating and reusing process water for other purposes like landscape irrigation. Firms should also consider using highly treated and reliable recycled wastewater where available.

CONCLUSION

Implementing the water efficiency measures listed above can limit your firm's supply exposure during future water shortages. Additionally, implementing cost effective water efficiency measures will save on water and sewer, as well as energy bills now. Finally, water efficiency measures are a way for your firm to demonstrate a "green industry" commitment by reducing water and energy demands while potentially reducing water pollution impacts as well.

Chapter 7

Sustainable Development And Green Landscaping

Leland Walmsley
Founder of everGREEN landscape architects
leland@everGREEN-sb.com

EDITOR'S COMMENTS:

This chapter has a wealth of information that considers environmental impacts that result from decisions made in the planning process. Leland has provided nuggets of interesting information throughout each section—some of this information you can even apply in your home!

INTRODUCTION

Many of these ideas are not new, merely recycled from an earlier era—a time before 10-ton vibrating compactors, before PVC irrigation and nitrate-based chemical fertilizers, pesticides and herbicides. Once upon a time, professionals practiced a holistic approach to landscape, architecture and horticulture. Those approaches can still be found in the pages of esoteric books and periodicals that are themselves recycling ideas for a new generation of thinkers—sustainably minded thinkers. Within the context of sustainable landscape architecture, there are five distinct areas that design, methods and materials may contribute to and should be seriously considered.

SUSTAINABLE SITES

Site Selection

Sustainable development should focus on urban areas with existing infrastructure (utilities, transportation, etc.) in order to protect open green space, preserve habitat and undeveloped natural resources.

Pick the Right Site

With so many empty structures and vacant lots in American cities, it is important that these once-occupied-now-abandoned sites be developed before outlying, previously undeveloped sites are built upon. Better to select an in-fill site where infrastructure (utilities, transportation, etc.) already exist; whereas outlying undeveloped areas will need new infrastructure at substantial cost and may be inappropriate due to environmental impacts. Inappropriate sites might include sites that result in the loss of prime farmland or public parkland, land within 50 feet of a water body, within 100 feet of a sensitive wetland, undeveloped land identified as habitat, land lower than 5 feet above 100-year flood elevation. Developing these sites may result in the loss of green space, habitat and natural resources forever.

Analyze Site

All sites are different and unique; and as such, it is critical to understand the larger ecological context to understand the opportunities and constraints of the site and to determine suitable development.

Identify and Work with Existing Site Features

Analyze, identify and consider: property boundaries, ridgelines, climate (rainfall, winds, sun angle, seasonal phenomena and microclimate), features that are unique, unusual, or special to the site, culturally significant landscapes, topography, hydrologic patterns (flow, flooding patterns and erosion), existing tree, woody and herbaceous plant species, indicate whether the existing plant is native or exotic, desirable or invasive, current land use patterns, historic land use, soils, location of the site within the local watershed and any impacts development may have, identification of brownfields or other environmental hazards, any environmentally sensitive features.

Decrease Exposure to Erosive Forces

Sun, wind and rain are the three most erosive forces. Site selection that considers these naturally erosive forces and the appropriate location and orientation of the structure along with the appropriate use of landscape can result in substantial energy savings and reduced maintenance costs (i.e. structure, paving, roof, etc.).

Correctly siting the structure based on the sun and seasonal solar orientation can result in huge energy savings. A structure with deciduous trees on the south and west sides will benefit from shading during summer months and reduce cooling costs. During winter when the trees are bare and the sun angle low, the structure will benefit from the sun penetration into the interior, added light, warmth and reduced costs. A deciduous vine (such as grape or wisteria) covering a trellis on a structures south side will achieve the same desirable effect. Trees and vines will partially shade the structure from the sun's harmful UV rays, resulting in less frequent need for paint. Even primeval man sited his dwellings on south slopes to take advantage of the low-angle winter sun that helps warm the structure.

Locating the structure below the top of a ridge or hilltop (within the wind shadow) means the structure will be spared frigid, penetrating winds. Together with evergreen trees placed on the north side to buffer against cold winter winds, means less draft and less heating. The structure will benefit from trees on the upwind (windward) side due to evapotranspiration. Evapotranspiration is the evaporation of water vapor through leaves. Much like a swamp cooler, evapotranspiration will cool the structure. (http://en.wikipedia.org/wiki/Evapotranspiration).

Development Density & Community Connectivity

A suitable site for development will consider existing adjacencies to the site including wetlands, wildlife corridors, urban-wildlife interface, other development, nearby goods and services, roads and infrastructure (utilities, transportation, etc.).

Integrate Community Values

By conducting community stakeholder meetings, they will be better understand how sustainable landscape design can support the culture, values, and aspirations of the local community. Within a 10 mile radius, look for and evaluate community input. Consider:

• Safety, crime statistics and security. Consider implementing "Crime Prevention Through Environmental Design" (CPTED)—using proven design measures to deter crime (http://www.cpted-watch.com/). Common crime prevention methods include excessive lighting with substantial equipment and energy costs. Crime statistics show that there are more energy-efficient ways of preventing crime than to emblazon a site in light. (http://en.wikipedia.org/wiki/Crime_prevention_through_environmental_design).

- Views, open space and environmental concerns addressed by local planning and ordinances.

- Public art.

- Outdoor and educational opportunities.

- Historic preservation.

- Ethnic cultures.

Cultural Setting

Encourage a sustainable landscape design that responds to the local community with respect to history, cultural resources, urban design, local land use regulations, transportation systems and public access. Consider these factors at various scales—from neighborhood to regional. Designers should preserve and enhance existing culture landscapes, and be familiar with:

- Regional master plans.

- City master plans.

- Transportation plans and routes.

- National Historic Register and Cultural Landscapes.

- Park and trail systems master plans.

- Stormwater management plans.

- Open space master plans.

Ecological Framework

It is important to restore or maintain the ecological pattern of the landscape with respect to:

- Topography—avoid disruption of landforms OR restore natural conditions with a sensitive approach to the site.

- Hydrology—do not alter drainage patterns, hydrology, existing riparian features OR restore to historic natural patterns, and avoid disruption to sensitive environmental features.

- Habitat—avoid disruption OR improve habitat patches and create habitat corridors. Remove all invasive, exotic plant and animal species, and institute a monitoring program to ensure successful removal.

- Education—provide public information about the ecological benefits of the site.

Develop Brownfield Sites and use Plants to Remediate Polluted Sites

Damaged and polluted sites formerly occupied by industry (brownfields), gas stations, etc., are great development sites that may be rehabilitated using natural processes known as bioremediation. Bioremediation utilizes trees, plants and microbes to extract, filter and bind up toxins in the soil, naturally; instead of excavating contaminated soil, transporting it far away (using fossil fuels) and disposing of it at a toxic waste site. Developing these sites reduces pressure to develop green space and habitat.

Trinity College (Hartford, CT) used mustard plants to extract lead from the soil on a site formerly occupied by a paint store.

In Fort Bragg, California, mushrooms are being used to mitigate dioxin contamination at a former lumber mill. (http://mendocoastcurrent. wordpress.com/2008/04/28/saddled-with-legacy-of-dioxin-fort-bragg-considers-an-odd-ally-the-mushroom/)

Poplar and willow trees are effective hydrocarbon cleaners, and new cultivars are being grown to clean up 300% more pollution.

Growing switch grass nurtures the life of the soil, sequesters carbon and helps prevent greenhouse gases, stores up to 27% of soil carbon and prevents it from escaping into the atmosphere. Unfortunately, high-phosphorus chemical fertilizers inhibit switch grasses' ability to do so.

(Also see "Create bioswales," below.)

Alternative Transportation

By understanding and promoting sustainable transportation, we greatly reduce pollution and the development of green space and habitat.

Public Transportation

Seriously consider access to public transportation, including: commuter rail, local bus routes, bike routes, and walkable sites.

Bicycle

Segregate pedestrians, automobiles and bicycles from one another. If possible, tie into existing neighborhood or regional bicycle routes. To encourage bicyclists, install secure bike racks (covered if possible), and changing rooms with shower and lockers. *For Example: West Coast Asset Management (Santa Barbara, CA) provides a bike rack, shower and changing room to contribute toward a LEED Platinum certification.*

Low-emission & Fuel-efficient Vehicles

Provide these vehicles or incentives to drive these vehicles to em-

ployees and provide preferred parking in the shade of trees and closest to the destination. *For Example: Allen Associates, a general contractor, (Santa Barbara, CA) has subsidized employees who turn in their gas guzzling trucks for high-mileage, alternative fuel and hybrid vehicles. Employees who use more expensive alternative fuel (biodiesel) are compensated for the difference in fuel costs. The company has a monthly drawing for $100 for anyone who carpools, takes alternative transportation or telecommutes.*
Also, they give alternatively-fueled vehicles preferential parking.

Parking

Reduce parking, pollution and the development impacts of single occupancy vehicles. Minimize available parking spaces. Encourage carpooling programs amongst employees, and give carpoolers incentives, such as preferred parking in the shade of trees and closest to the destination. Consider sharing parking with adjacent structures that have different hours of operation.

Site Development—Protect or Restore Habitat

The conservation of existing natural areas and restoration of damaged areas provides important habitat and biodiversity. To do so, limit construction disturbance to the immediate perimeter of the structure and area of hardscape (roadway, sidewalks, patios, parking and utility locations).

Maintain Soil and Plant Health

Soil health alone can result in a landscape's success or failure. For this reason, we must maintain and improve soil quality, continue to improve the soil's ability to function as a vital living system, and contribute to the health of the overall ecosystem. However, current engineering standards for construction and paving typically result in very poor soil quality that is typically compacted to 95%. Compaction restricts the health and vigor of most all trees and plants and hinders stormwater infiltration. High compaction, inorganic debris and pollutants all contribute to the premature death and reduced environmental benefits of urban trees.

Newly engineered "support paving" provides structural support for paving and the load it carries while soil below remains compaction-free and tree roots grow healthy. Furthermore, high quality soils and expanding root zones of healthy trees further retain stormwater, reducing the possibility of flooding.

Improve Hydrology

With the increase of paved areas, stormwater runoff has increased. Increased runoff volumes can be offset by creating bioswales and rain ponds. Bioswales and rain ponds are variations on depressed, planted areas in the landscape that create restored habitat. Runoff is directed toward them, pollutants are cleansed by plants, and water percolates to recharge the water table. Bioswales are not intended to have standing water. Stormwater runoff is quickly absorbed by the deep rooted plants without a chance for mosquito to breed. (See "Reduce stormwater quantity" below.)

Not long ago, wetlands were considered vermin filled swamps that were filled in with excess construction debris and earth. Now, paving is being removed and wetlands are being restored; not only for the benefit of the creatures that live there but also for the valuable purpose they serve, mitigating pollutants and minimizing flooding. Bioswales and rain ponds are not the complete solution, but they do help mitigate increased runoff volume resulting from impervious paving associated with urban development.

Use Native Plants

Determine plant and habitat selection based on the site's soil. Revive degraded soils with a rich and varied supply of organic matter. Consider the potential of unamended soils to support adapted, desirable plant life. Create a diverse landscape and avoid monocultures. Protect the habitat of soil organisms by specifying: reduced tillage, minimal compaction, minimal the use of pesticides, herbicides and fertilizers, improved drainage and maximized plant coverage.

Encourage the use or protection of regional plant with an emphasis on natives to the site. Do not use exotic/introduced plant that is invasive and causes degradation of local ecosystems, especially riparian areas. Natives are better suited to local environmental and climatic conditions, less apt to need supplemental irrigation and support native wildlife, such as local insects and pollinators that help sustain populations up the food chain.

Encourage Wildlife

Limit construction disruption and reduce threats to desirable wildlife and associated habitat. The landscape design should increase, create, or restore desired habitat or wildlife presence at the site. Remove undesirable habitat or invasive exotic species.

Site Development—Maximize Open Space

Sustainable development should provide a high ratio of open space to development footprint to promote biodiversity. Create vegetated areas adjacent to the structure, including green roofs and shared open space with an adjacent structure, beneficial to both developments.

By maximizing open space and creating a view of healthy vegetation is synonymous with healthy humans and patient recovery. Prevailing hospital design includes open space and gardens due to evidence showing 80% faster patient recovery when patients have a garden view. (http://www.gghc.org)

Pollution Prevention during Construction

In pursuit of a sustainable structure, great care should be taken to limit disruption from construction activities and avoid pollution of natural ecosystems.

Provide Trash and Recycling Receptacles On-site

Discuss with contractors the importance of proper disposal during demolition and construction. Specify where and how discards are to be dealt with.

Reduce Soil Compaction

Driving a vehicle on bare earth and parking in the shade of a tree may seem harmless, but easily results in long-term damage—soil compaction. The active root zone of trees includes the drip line of the canopy and 5-10 feet beyond. Typically, undisturbed soil has 15-20% oxygen by volume. Like human beings, healthy roots require oxygen. Oxygen resides in soil voids and is critical to the health of all trees and plants. Soil compaction reduces oxygen in the soil.

Limit Equipment Use

Driving over tree roots even briefly may result in the slow decline and the possible death of the tree. To help protect against soil compaction, always designate parking areas away from existing trees and plants and areas intended for future plantings. Use temporary orange fencing to protect active root zones from automobiles and equipment.

To further protect trees, wrap low-hanging branches with padding and brightly colored sheathing to avoid damage by equipment. Equipment damage to branches might include a truck or forklift carrying materials. Damage to equipment from low-hanging branches can also be substantial.

Reduce Soil Contamination

Soil contamination most often occurs when vehicles park within the root zone of an existing tree. A truck with a leaky oil pan can kill a tree after just one day. To avoid soil contamination, adopt the same measures to deal with soil compaction and soil contamination.

Another shortcoming of construction is the disregard for plants by many trades. Often trades (drywall, paint, stucco, etc.) clean their tools and dump the washout on an open area thinking that because nothing is growing there at the time that it is ok. Cleaning and dumping a cement mixer's washout onto soil will seriously affect pH, leading to toxic soil that can no longer support plant life without major mitigation including soil removal, disposal and replacement; all at substantial cost. To avoid this, designate washout areas where trades can clean up safely away from existing and proposed planting areas. Specify a lined 55-gallon drum is provided for trades to dispose their washouts and clearly explain washout procedures to all.

In the event of a spill, establish contamination procedures and clearly explain those procedures to all trades.

Preserve Topsoil

Soil primarily erodes by wind as it did during the dust bowl. In addition to losing topsoil, wind erosion also results in air pollution. "Impact erosion" occurs when rain strikes bare soil and washes it away. On a slope, "impact erosion" can be quick and devastating. There are many ways of preventing soil erosion. Coir wattles made from coconut husks, silt fences, straw bales, and a layer of mulch are all very effective at minimizing topsoil erosion. To reduce erosion (and air pollution), spread 4-6" of mulch (wood chips). Use mulch from on-site trees and plants or from the local vicinity. That way, if pathogens, fungi, etc. are present, trees and plants may have already built up a resistance. Whereas, if imported mulch inadvertently introduces a new pest, existing trees and plants may not survive the newly introduced threat. In some counties of California, mulch/wood products may not cross county lines due to sudden oak death syndrome. (http://nature.berkeley.edu/comtf/)

Whether coir wattles, silt fences, straw bales or mulch, all municipalities require some form of construction erosion control under the US

EPA's National Pollutant Discharge Elimination System (NPDES) which requires all point sources discharging pollutants into waters of the United States to obtain a permit.

So why is topsoil so important? "In 150 years since plows first broke the plains, Iowa has lost one-half of its topsoil, and the rate shows no sign of slackening."[1]. In most places, topsoil is only 9" deep. Topsoil is the nutrient-rich layer of earth where 90% of the soil life resides. Yes, soil is alive. A single teaspoon of soil contains four billion life-forms—a perfect balance of nutrients needed to properly feed plants. Soil microbes perform specific functions to create a healthy growing environment. Over millennia, these same microbes made the fertile soil we have today. Understanding soil biology is the key to sustainable landscapes and healthy soil; and healthy soil means a reduced need for chemicals. And once soil disappears into the wind or washes into waterways, it takes millennia to replace the topsoil that has been lost.

The benefits of healthy soil include improved drainage, soil aeration and less compaction, less erosion due to healthier root systems, better water absorption resulting in less irrigation, reduced plant disease and pests, decomposition of toxins, reduced groundwater pollution due to living soil biology consuming pollutants.

Reduce Stormwater Quantity.

According to research by the US EPA in 2003, a typical city block generates 5 times as much stormwater runoff as a wooded lot of the same size. Due to the built environment and little pervious open space, less rainwater percolates into the earth. With less water percolating down, water tables are dropping and stormwater runoff volumes above ground are increasing. Prevailing civil engineering strategies are to drain the water away from the human environment as soon as possible. To do this, water is directed to concrete-lined streets, channels and pipes—once again increasing imperviousness and decreased percolation, and further urban flooding. Without permeable surfaces such as streams, runoff volumes increase in a vicious circle whereupon concrete channelizing begets more channelizing. Hence cities are seeing more urban flooding in areas that never flooded before. And of course, more paving and concrete means less habitat.

However, there are many ways in which we can both reduce stormwater quantity and increase groundwater recharge by means of bioswales/ rain ponds, porous paving, infiltrators under parking lots and curb-side constructed wetlands.

Harvest Rainwater

Rainwater is perhaps the most valuable, clean, nutrient rich water we can hope for. Rainwater leeches salts and minerals from the soil that accumulate overtime and ultimately prove toxic to plants. Rainwater has balanced pH and is the best thing we can put on out plants, but prevailing thought is to get rid of rainwater runoff for fear that it will result in flooding.

Using a cistern, rainwater can be captured and stored to reduce stormwater runoff quantity. Captured water may be used for irrigation or to flush toilets. During dry months, collect condensate from air conditioning or runoff from water used to clean photovoltaic panels.

Use Porous Paving

From 2000 to 2005, California paved over an area equal to 1-1/2 Rhode Islands. As a result, rain water is unable to replenish ground water. Indigenous tree roots typically dip into this ground water especially during summer months. But if the ground water level drops, it can mean death to many trees. Tucson, Arizona has seen its water table drop from 20 feet below the surface in 1950 to over 250 feet below the surface in 2007. Parking lot surfaces now include structural geo-grid (grassy pavers) planted with grass or groundcovers or merely gravel. The geo-grid supports the weight of a emergency vehicles, and reduces stormwater runoff which percolates. Ground water is replenished, contributing to the health of trees. Trees in turn evapotranspire which cools the urban environment and minimizes "Heat Island Effect—Non-Roof" (see below). Many parking lots implement planted parking on the periphery, such as overflow parking, or the less often used handicapped parking.

A relatively new product, porous concrete, allows for water penetration, but the porousness diminishes as "fines" fill the permeable voids.

For traditional (impervious) asphalt paving, stormwater runoff may be directed to infiltrators that are placed under traditional parking lots and work much like a leach field capturing runoff and allowing for slow percolation.

Construct Bioswales/Rain Ponds

Once again, bioswales/rain ponds can lessen the affects of runoff. The SEA (Street Edge Alternative) Streets in Seattle is an example of successfully reversing the urban runoff dilemma. Previously, the creeks and sounds around Seattle were seeing a dramatic decrease in salmon due to polluted urban runoff. By eliminating one street parking space per block, rain ponds were created in depressions. Street runoff was diverted into these depressed plant areas via curb cuts. There, the plants slowed the flow and captured urban pollutants, including heavy metals, cigarette butts, petrochemicals, etc. and allowed runoff to soak in and recharge the water table. A University of Washington study showed some SEA Street areas reduced runoff by over 90% in the winter and 100% in the summer. Now, the salmon are back in abundance. The SEA Streets areas initially met with opposition due to a few less street parking spaces. But real estate along the SEA streets sells well above average market value. Businesses report better foot traffic and sales than similar businesses on streets without. (http://www.google.com/search?hl=en&q=SEA+street s)

Bioswales/rain ponds can just as easily be implemented into a parking lot. Downspouts can be directed toward a bioswale/rain pond. Locate the bioswale/rain pond down-slope from the structure to avoid mold and mildew issues. Generally, the size of the bioswale/rain pond should be 10% of the area that drains to it.

In 2005, Kansas City voters approved a $500 million bond issue funding a new and improved water infrastructure. Though it will be years until all of the necessary work is completed, Kansas City officials started the "10,000 Rain Gardens" project to help manage stormwater immediately. (http://www.rainkc.com) Officials believe there are already more than 1,000 rain gardens in place. The new program will give property owners an incentive to register their rain gardens so the count is accurate. The program's goal is 10,000 rain gardens in five years, by 2010.

Green Roof

These planted roofs also reduce stormwater quantity as they absorb stormwater to a point.
(See "Improve Stormwater Quality," below)

Improve Stormwater Quality.

"First flush" occurs during the first 20 minutes of the first substantial rainfall of the season and has the highest concentration of pollutants, 10 times higher than the concentration at the end of a storm event.

In California (and many other areas), the first rainfall occurs over distinct periods. A long dry period, from April to October allows contaminants to build up, creating a larger discharge when it first rains again. The initial surge of contaminated stormwater ("first flush") includes: litter, fecal coliform, bacteria, viruses, heavy metals, petrochemicals & hydrocarbons from roofs, streets, parking lots and an accumulation of lawn pesticides, herbicides, and chemical fertilizers. This toxic soup contributes to algae bloom which depletes oxygen from lakes and streams, kills fish and riparian life, and taints the food chain.

If we filter the "first flush," we effectively capture approximately 80% of the contaminants. Filtration can be achieved with relatively inexpensive manufactured filters or naturally using plant materials. A well conceived landscape uses plants to naturally (and inexpensively) filter out pollution before the stormwater runs into lakes or streams. Oil, chemicals and other pollutants are filtered and stopped by soil particles and taken up by roots. As mentioned above, poplar trees are being breed to improve upon the tree's natural ability to uptake the worst pollutants.

Of course, pervious surfaces, bioswales/rain ponds, and green roofs improve stormwater quality using plants to naturally filter out pollutants.

Reduce Heat Island Effect

A result of development is increased urban temperatures. Increased radiant temperatures are caused by dark colored hardscape (asphalt) and buildings that effective act as heat sinks, affecting humans and wildlife habitat alike. To lessen the affect, shade constructed surfaces—structures and hardscape.

Plant Trees

The urban forest has been shown to reduce Heat Island Effect. In Tucson, Arizona, tree canopy has lessened summer temperatures by an average of 9 degrees, mostly by shading black asphalt which can retain daytime heat throughout the evening (See "Use plant material to decrease energy needs," below).

Green Walls

Go vertical and plant a green wall on the east, south and/or west side of the structure. Once again, a green wall will shade the structure and help reduce temperature. There are two main categories in green walls: "green facades" and "living walls."

"Green facades" are usually climbing vines growing either direct-ly on the wall or on specially design structures, such as greenscreen. (http://www.greenscreen.com)

"Living walls" are bit more involved and more impactful. Not many other sustainable methods or materials identify a "green" structure to the public like a living wall. "Living walls" are made up of pre-vegetated pan-els attached to a wall. These walls can be put up anywhere as long as there is access to light (artificial or sun), water, and fertilizer. The walls are watered from the top, and the watering and fertilizing are usually auto-mated. (http://images.google.com/images?hl=en&q=%22living%20wall s%22&um=1&ie=UTF-8&sa=N&tab=wi) Indoor, living walls can contrib-ute to improved air quality (see "Improve Air Quality," below).

Reduce Heat Island Effect—Roof

A traditional roof represents a heat sink during summer months, day and night due to materials (asphalt and concrete) and the lack of veg-etation in urban areas. To address Heat Island Effect on the roof, plant a green roof. With green roofs, roof temperatures during summer have been reduced from 150 to 90 degrees (http://en.wikipedia.org/wiki/Heat_is-land_effect).

Green Roof

Not new, green roofs have been around since the Vikings who used them to insulate their structures from the bitter cold. Today's green roofs insulate from winter cold and summer heat. With added insulation, HVAC, all related equipment and ducting can be reduced by 1/3; reducing equip-ment and operating costs substantially.

Without exposure to the UV, a roof's life may be extended up to 50 years. Furthermore, green roof plants absorb rainfall, reducing stormwa-ter quantity and improving stormwater quality. One roof may not seem like much, but Chicago currently has accumulated more than 4 million square feet of green roof. In fact, a public project must have a green roof in order to receive funding.

In Dearborn, Michigan, Ford Motor Company built the largest green roof in the world atop its 10 acre truck plant. An unintended result of the Ford green roof is restored habitat. Birds are nesting free of predators (cats, foxes, snakes, etc.) And "pioneer plants" from bird droppings and wind are taking root, creating further biodiversity. (http://www.google.com/search?hl=en&q=Ford+green+roof)

(For more on green roofs - http://www.greenroofs.org/)

Reduce Light Pollution

According to the International Dark-Sky Association, light pollution is any adverse effect of artificial light, including sky glow, glare, light trespass, light clutter, decreased visibility at night, and unnecessary energy waste. (http://www.darksky.org/mc/page.do)

Artificial light at night has been shown to affect the mating, migration, and predation behaviors of many different species. Specified periods of light and dark are essential for good health for most life on Earth, including human beings. Recently, health researchers have established that exposure to artificial light at night reduces the human body's production of melatonin. Reduced levels of melatonin promote growth of breast tumors in women and may similarly affect other cancers, including prostate cancer.

To reduce the impact of development:

• Select low-glare lighting equipment.

• Use outdoor light at night only when and where it is needed and at appropriate lighting levels.

• Use fully shielded, efficient fixtures aimed directly at the ground, and avoid up-lighting.

• Aim facade/architectural lighting down from the top, if at all possible; otherwise, make certain that any up-light does not escape the lines of the structure.

• Shield landscape and security lighting so that the majority of light hits the target is shielded from normal viewing angles and does not cause glare.

• Do not over light an area. Excessive lighting does not prevent crime. However, crime statistics show that there are more energy-efficient ways to prevent crime than to emblazon a site in light. Excessive use of light merely adds to equipment and energy costs. Lower lighting levels and better uniformity account for improved safety and security.

• Incorporate timers and sensors to shut off lights when not needed. Landscape and facade lighting should be turned off after midnight or earlier. Many parking lot lights can also be turned off after hours.

Look for the International Dark-Sky Association (IDA) "Fixture Seal of Approval" for dark sky friendly fixtures. (http://www.nextrionet.com/mc/page.do?sitePageId=56423)

WATER EFFICIENCY

Water Efficient Landscaping

Society pays a heavy price for potable drinking water. 3% of U.S. public energy demand is used to move water. In California, that number jumps to 20% with 8% occurring at the end user site (heating water, reverse osmosis, pool pumps, etc)[2]. Therefore, to save water is to save energy. Half of the potable drinking water is used for non-potable applications, landscaping and agriculture. Due to the unnecessary cost associated with safe, potable water, a sustainable site might best limit or eliminate the use of potable water for non-potable purposes (i.e. irrigation and toilets).

Use Plants with Little or No Water Needs

In order to minimize water requirements, use regional native plants based on site location. Once established, these plants need little to no water depending instead on naturally occurring rainfall. Additionally, these native plants may provide much-needed food for migratory birds to make up for habitat loss. (See "Use Native Plants," above)

Harvest Rainwater

(See "Reduce Stormwater Quantity," above)

Non-potable Municipal Water

Plants have no need for potable drinking water. Non-potable water will do. Check with the local water agency to see if there is a non-potable source near the site in question. If there is a public park or golf course nearby, then there may be a non-potable water supply for purposes of irrigation, available at a much reduced rate.

Spread Mulch

In addition to preserving topsoil and reducing erosion, mulch keeps soil cool in summer and warm in winter will facilitate greater and deeper root growth. Mulch helps maintain soil moisture longer further minimizing irrigation needs and thwarts the weeds that compete with the desirable plants for water. Mulch can easily reduce water needs in half.

Add a Weather-based Irrigation Controller

Landscape plants can drink up half of the domestic water use. Weather-based irrigation controllers create a site-specific irrigation schedule for any landscape. Menu choices include: sprinkler type, slope angle, sun exposure, soil type, geographic location and plant material types. Minute by minute, weather stations measure wind, rain, temperature, relative humidity, and solar radiation and compute the amount of water vapor given off by plants (evapotranspiration) to the .01 inch. In real time, sensors recalculate water needs, adjusting to deliver irrigation with exacting precision to save water and money. Weather stations can be wireless, solar powered and easy to install.

Sites with vast landscaping, i.e. theme parks, parks, R&D campuses, golf courses, etc, are saving hundreds of thousands of dollars per year by irrigating exactly what the plants require. There is no guess work and with the correct amount of water, plants are healthier, resulting in less maintenance.

Hydrozone Plant Areas

Irrigation should be zoned by plants with similar water requirements, similar growing conditions (sun, shade, hillside, etc.). Lawn irrigation should be separated from plant beds as they have vastly different water requirements. Different sprinkler types (i.e. drip and pop-up sprays) should not be mixed as they put out different volumes of water.

Maintain Irrigation Regularly

It is critical that irrigation systems should be inspected monthly for clogs, proper sprinkler/drip function and broken pipes. The delivery time of day should be set for morning to minimize evaporation and potential fungal growth. The controller schedule should be changed to accommodate seasonal changes, and to adjust for the maturation of new plants.

Use Efficient Irrigation

Drip irrigation has come a long way, primarily perfected in Israel where water has always been scarce. Drip irrigation will substantially reduce water requirements as very little evaporation will occur, water is placed directly on plants and little water wasted on non-planted areas.

Replace the Lawn

Considering water and maintenance costs over-time, the traditional lawn is the most time-consuming, energy-consuming and expensive plant per square foot in the garden. Consider low/no water use grass alternatives. In California, alternatives include *Bouteloua gracilis* (Buffalo grass), *Dymondia margaretae* (no common name), *Juniperus conferta* (Shoreline juniper) and *Carex cultivars* (Sedge, native to all 50 states). Once established, these alternatives need little to no mowing, irrigation, chemical fertilizer, herbicides or pesticides.

Many cities are offering rebates for lawn removal. The Southern Nevada Water Authority (primarily Las Vegas) offers a $2 rebate per square foot for lawn removal up to 1,500 square feet, and a $1 rebate for additional removal thereafter. Through 2006, nearly 80 million square feet of thirsty lawn had been removed. In Las Vegas, a typical 1,500 square foot lawn conversion saves more than 80,000 gallons of water per year. (http://www.snwa.com/assets/pdf/ar_2006_conservation.pdf) Furthermore, less lawn means less water, less chemicals, and less maintenance. The absence of mowing, edging and blowing means less fossil fuel use.

By replacing the traditional lawn substantial costs can be saved.

Implement Wastewater Innovations
Use Greywater

Clothes washers are the second highest water usage device behind toilets. Front loading washers use up to 18,000 gallons per year for an average family. With the right detergent, wastewater from a washing machine is just fine for the purposes of irrigation. Except for toilet and kitchen disposal/dishwasher wastewater (blackwater), all other washwater (greywater) can be used a second time—on the landscape. Greywater reduces the need for potable water on plants. Greywater is perfect for bioswales, native plants and trees. And in fact, most of the world's greywater is used in that way. Here in the United States, the Uniform Building Code requires greywater be treated and managed similar to wastewater in the interest of public health. Most safety officers frown on greywater altogether, but greywater can be legally treated and dispersed much like a wastewater in a septic tank. (http://en.wikipedia.org/wiki/Greywater)

Another form of greywater that helps recharge ground water includes outdoor drinking fountain runoff. Outdoor drinking fountains should be built atop a gravel sump. Rather than piping the relatively clean runoff to the sewer, the gravel sump allows for excess water to percolate and recharge groundwater without puddling at the surface.

ENERGY & ATMOSPHERE

Create a Maintenance Program

Encourage practices for maintaining and managing the overall site (grounds and structures) to minimize environmental harm and to preserve ecological integrity and overall health of the people, plants and animals using the site. Encourage site design that enables environmentally sensitive and low-impact maintenance and management practices. Maintenance considerations include: equipment, plants, pest control, green waste reduction, stormwater filters, irrigation, fertilizer use, pesticide use, snow removal (where applicable), outdoor structures, furniture maintenance, maintenance of structural exteriors, water features and outdoor drinking fountains, hardscape maintenance, lighting fixtures, signage, sculpture and artwork.

Reduce Energy needs.
Consider a Wind Turbine

Vertical axis turbines are lower profile than horizontal axis (windmills), require less wind, less height, thus incorporated into the architecture and can even be quite artistic. Seeing an artistic "green" feature helps raise awareness and educates the public as to new ways of "greening" a structure.

(http://images.google.com/images?hl=en&q=%22vertical%20axis %20wind%20turbine%22&um=1&ie=UTF-8&sa=N&tab=wi)

As previously discussed ("Water Efficient Landscaping"), moving water amounts to substantial energy use. Less irrigation results in substantially less energy consumption.

Optimize energy savings in the landscape.
Look to Emerging Technology

There are many emerging technologies, including: solar irrigation controllers, solar weather stations, and solar lamps. LED lamps replace traditional incandescent and fluorescent bulbs, provide the same lumens while using only 2-3 watts, or $1/10^{th}$ the energy of traditional lamps. With LED lamps, the size of costly copper wire is reduced, fewer expensive transformers are needed, energy consumption reduced by 90%, and lamps live 50 times longer resulting in less maintenance.

Some of the emerging technologies on the horizon include bioluminiscence—converting chemical energy into light energy, displayed by mushrooms, jellyfish and insects that glow in the dark. By understanding how nature's bioluminescence works, we may incorporate that into our path and street lights. (http://en.wikipedia.org/wiki/Bioluminiscense)

Use Plant Material to Decrease Energy Needs

Place structures, outdoor seating and parking with trees, plants and vines in mind. Vegetation can help greatly reduce the affects of thermal mass and radiant heat associated with Heat Island Effect (See "Reduce Heat Island Effect—Non-Roof," above).

Plant Trees

"If we lose forests, we lose the fight against climate change," declared 300 scientists at the December 2007 climate conference in Bali. The U.N. estimates that 60 acres of tropical rainforest are cut down every minute (or nearly 31 million acres per year), up from 50 acres per minute a decade ago.[3] Furthermore, the burning of trees that comes from deforestation sends more carbon dioxide into the atmosphere than all of the world's planes, trains, trucks and automobiles. Whereas, healthy forests absorb (sequester) carbon dioxide.

Trees are a low-tech solution to energy problems and are the easiest way to cool a structure. Trees help reduce energy costs in many ways. Deciduous trees on the east, west and south sides shade a structure providing relief from the sun during summer months, reducing the need for air conditioning. In winter, when tree branches are bare, winter sun warms the structure. Placed appropriately, conifers deflect winter winds upward which reduce heat loss and cold air infiltration, minimizing heating fuel costs. Trees reduce the quantity of stormwater runoff that contributes to flooding, and improve the quality of runoff cleaning the water. A single mature tree can absorb 5000 gallons of stormwater runoff, sequester one ton of carbon dioxide from the air per year that leads to global warming, and leaves filter and clean the air. Trees improve the mental well-being of building occupants. Hospital patients have been shown to recover more quickly from surgery when their hospital room offered a view of trees. Wood is a renewable source of fuel which lowers the demand for fossil fuel energy.

In Australia, a program called "CO_2 Australia" pays land owners to plant the *Eucalyptus polybractea* (Mallee Tree) which is long lived, pest resistant and drought tolerant to remove and sequester carbon dioxide and help reduce global warming. (http://co2australia.com.au/)

MATERIALS & RESOURCES

Avoid using materials proven to be harmful to human and animal health, water, soil, or air quality.

Storage & Collection of Recyclables

Always provide trash and recycling receptacles. Studies show that all people (irrespective of ethnicity, social class, gender, age, mental capacity, etc.) will use trash receptacles if given the opportunity. It is largely the absence or poor placement of receptacles that leads to pollution.

Reuse Materials

Compost

Compost can replace all of the chemical fertilizers, pesticides, fungicides, and some herbicides that we've been using in the landscape. By replacing these chemicals, toxic residues are reduced, such as nitrate poisoning in groundwater, pesticide residue on lawns and salt buildup in soil.

Compost is black gold. Why pay to export organic garbage rich in nutrients off-site then pay again to import chemical nutrients on-site? All compost needs is air, sun and moisture. If compost is too adventurous, hire an all-natural landscape service that will inoculate and feed the plants with compost (or worm) tea. It wasn't so long ago that this natural way was the only way gardening was done.

A composting mower leaves the nitrogen-rich grass trimmings to breakdown, provides nutrients, resulting in a need for fewer chemicals. Fewer chemicals mean less sickness and absenteeism, greater worker satisfaction and better productivity.

Coffee Grounds

Ever consider how much coffee grounds contribute to the waste stream? Most every office has a coffee machine using as much as a quarter pound of grounds per day, five days a week, and 50 weeks per year. Why not benefit from those grounds as compost and reduce the waste stream? (http://oregonstate.edu/dept/ncs/newsarch/2008/Jul08/coffee.html)

"Coffee logs" reuse coffee grounds to make fireplace logs. They have substantially lower air emissions compared to wood, produce 25% more energy, divert millions of pounds of coffee waste from landfill every year, save trees, use 100% recycled packaging and reuse natural vegetable wax, without chemicals. (http://www.java-log.com/)

Divert waste from disposal.
Reduce
By mulching compostable scraps including paper, we greatly reduce the amount of waste going to our landfill and the fossil fuels needed to transport.

Purchase goods, services and food from local vendors, reducing the fossil fuels needed to transport.

If trees produce edible fruit that goes uneaten, donate spare food to a local food harvest program that picks and distributes fruit to seniors and low-income families in the community.

Donate unwanted plants and site furnishings to a local charity rather than discarding as garbage.

Reuse & Recycle Materials
Add Fly Ash to Concrete Mix
Fly ash has been used since the Romans. It is resistant to salt and improves the strength and durability of concrete. It is the left-over reside from coal plants. By using fly ash, the need for mined cement is lessened. Furthermore, cement off-gases carbon dioxide, whereas fly ash does not. By using fly ash in concrete, you minimize your project's carbon footprint. (http://en.wikipedia.org/wiki/Fly_ash)

Consider "urbanite." Reuse broken pieces of concrete and asphalt (also called "urbanite") stacked as low decorative garden walls, pavers and pathways. In so doing, you reduce your project's contribution to land-fill, reduce fossil fuel used to transport waste, and decrease the need for new materials; all resulting in cost savings.

Greywater
(See "Use Greywater," above.)

Use Composite Woods
Composite woods are made of recycled plastic and wood pulp. Most commonly they are used for raised decks, pier decking, bench slats, over-

heads, parking lot tire stops, lawn headerboard, playground equipment, fencing and fence posts. One example is Trex™, but there are many wood composite products in the marketplace; and some are indistinguishable from real wood, except that wood composites don't need water sealers, don't splinter and last much longer than wood.

Use Asphalt Additives
Plastic music CDs, rubber and glass are common asphalt additives that reduce the contribution to landfill. The city of Santa Monica, California recycles tires, making sidewalks that flex rather than crack and heave if a tree root should swell.

Recycled Glass
In Broward County, Florida, ground glass bottles are being turned back into sand for local beaches rather than pumping sand from offshore sources that could damage reefs.[4] Recycled glass can also be used as a decorative groundcover.

Use Regional Materials
Use Local Materials
Seek products made within 500 miles. As fossil fuel becomes more expensive and transportation costs go up, local materials are much more affordable and carbon emissions are reduced.

In Santa Barbara, California, a local miller uses on-site trees that need to be removed. Rather than waste, he provides his clients with finished lumber used for floors and cabinets—all produced on-site.

Use Rapidly Renewable Materials
Over 100 million structures around the world are made of bamboo. It grows substantially faster than most hardwood trees. (http://www.bambooliving.com/construction.html)

Wood is a rapidly renewable material, but logging practices should be considered as well. Consider "Composite Woods" (above).

Use Certified Wood
If you must use wood, be sure to specify certified wood, such as the Forest Stewardship Council (FSC) and Sustainable Forestry Initiative. Trees are planted as trees are harvested and habitat renewal is considered as a consequence of logging, and certified by a third party.

IMPROVE INDOOR ENVIRONMENTAL QUALITY

Create Smoking Areas

Despite personal feelings about smoking, don't underestimate the importance of a well-designed smoking area. Smoking areas should not be near an entrance, but readily identifiable to users, and that area should always have a butt receptacle. If smokers are not provided with a smoking area, they will choose one by default; probably in an out-of-the-way place without a cigarette receptacle, which will ultimately result in butts littering the planter. (See "Storage & Collection of Recyclables," above). With time and stormwater runoff, those butts will make their way into the watershed unless properly disposed of.

Use Low-emitting Materials

Minimize the use of PVC. Vinyl chloride, a byproduct of polyvinyl chloride is one of the most carcinogenic compounds known, and yet one of the most widely used products around the world, having replaced traditional building materials such as wood, concrete, and clay. Despite all this, PVC is the current industry standard for irrigation with few alternatives. HDPE (high density polyethylene) is the only widely alternative suitable for irrigation mainline and drainage pipe. Sadly, it is not widely available and many contractor's are reticent to use it as PVC and HDPE are handled differently and with different solvents and glues, meaning twice the parts and equipment needed on hand. Ease and repetition are eliminated, and thought is required to the dismay of the contractor.

Avoid Using Treated Wood Waste (TWW)

TWW is commonly used for ground or water contact applications. TWW commonly includes arsenic, chromium, copper and pentachlorophenol. These chemicals are known to be toxic or carcinogenic. Harmful exposure includes contact with skin, inhalation from smoke or sawdust. Instead use composite woods (See "Use composite woods," above).

Avoid using stucco and paint finishes containing volatile organic compounds (VOCs). (http://en.wikipedia.org/wiki/Volatile_Organic_Compounds) Better to use water-based finishes with low or no VOCs. There are many brands now available.

Improve Air Quality

The U.S. Environmental Protection Agency estimates that Americans apply at least 90 million pounds of pesticides and herbicides to their lawns and gardens every year. Use of pesticides has risen steadily since 1998, and now represents the only growth sector of the U.S. pesticide market. American homeowners use up to 10 times more pesticides per acre of lawn than farmers use on an acre of crops. A study by the University of Oregon showed that pesticides and contaminants in soil reduce the yield of soybeans by one-third.[5] In fact, lawn and garden care is more of a danger to our health and the environment than conventional chemical-laden agriculture. Pesticides are applied more intensively for lawn care, where children, who are more vulnerable than adults, play. Much of these chemicals are subsequently tracked indoors, degrading indoor air quality.

Eliminate the Use of Chemical Fertilizers, Pesticides, Herbicides, Fungicides, etc.

Landscape chemicals are a byproduct of World War II's nitrate based weapons and byproducts of the petrochemical industry. With them, we toxicity the soil making it void of all life, microorganisms, nematodes, worms, etc. We must remember soil is alive. A teaspoon of soil has up to 4 billion microorganisms. There are just as many good microorganisms as bad, and plants depend on them as an integral part of the web of life. Our chemicals are not selective and kill most all microorganisms, good and bad, leaving the soil inert and dead. Frequently chemicals are over-used and plants can not uptake them all. With rain, the chemicals (mostly phosphorus and nitrogen) wash into the watershed where they create algae bloom. Algae depletes the available oxygen in the water and suffocates the fish, creating "dead zones."

The standard "weed and feed" combine chemical fertilizer and herbicide in the same bag, making the situation worse by again over-using costly products that may not even be necessary. One of the most common herbicides in "weed and feed" products, a chemical called 2,4-D, has been linked to human health problems, including an increased risk for non-Hodgkin's lymphoma. Of the 30 most commonly used lawn pesticides, 16 pose serious hazards to birds, 24 are toxic to fish and aquatic organisms, and 11 have adverse effects on bees, including bee colony collapse. Herbicides and pesticides from lawns also get into our water supply.

A study of 12 urban streams in the Seattle metro area found 2,4-D in every stream and 23 different types of pesticides, including five that were present in concentrations high enough to kill aquatic life. The researchers found a correlation between the pesticides polluting the sampled streams and the sales of lawn and garden chemicals from local retailers.

A recent report by the Toxics Action Center on ChemLawn—the largest provider of lawn care services in the U.S. uncovered that over 40% of the chemicals of their consumer product range contain ingredients banned in other countries, and all of the products in their lineup not only pose a threat to human health, but to water supplies, aquatic organisms, and non-target insects according to the Pesticide Action Network North America (www.panna.org).

The U.S. Geological Survey's National Water-Quality Assessment Program[6] found that 90 percent of the stream and fish samples surveyed contained at least one pesticide. To this end, a study of indoor air pollutants found 2,4-D in 63 percent of homes. A different study demonstrated that levels of 2,4-D in indoor air and on indoor surfaces increased after it was applied on lawns. The *Journal of the National Cancer Institute* found that exposure to garden pesticides can increase the risk of childhood leukemia almost sevenfold. Researchers at the Harvard School of Public Health found that frequent exposure to pesticides increased the incidence of Parkinson's disease by 70 percent.

Of the 30 commonly used lawn pesticides, 19 are carcinogens, 13 linked with birth defects, 21 with reproductive effects, 15 with neurotoxicity, 26 with liver or kidney damage, 27 are irritants, and 11 can disrupt the hormone system. Pregnant women, infants and children, the aged and the chronically ill are at greatest risk from pesticide exposure. Pets too are regularly poisoned.

A study revealed that exposure to lawns treated with herbicides four or more times a year doubled a dog's risk of canine lymphoma, while the Journal of the American Veterinary Medical Association reported that, when exposed to chemically treated lawns, some breeds of dogs were four to seven times more likely to suffer from bladder cancer. Cancer risk is significantly greater for domestic pets in homes where pesticides are regularly applied to lawns, tracked in on paws and feet, becoming indoor dust and breathed by pets and humans alike.[7]

What's worse, the EPA only requires fertilizer and pesticide manufacturers to list "active" ingredients on a product's label. The manufacturers are not legally required to disclose the inert ingredients, which can include harmful amounts of heavy metals. Inert ingredients in a lawn chemical may not kill the weeds, but there is no guarantee that they will be nontoxic to people or pets.[8]

Go Organic

Better to use organic, all-natural fertilizers, fungicides, herbicides and pesticides which target specific problems and are not harmful people or pets. Prior to WWII, gardeners and farmers did just that. Pathogens, invasive plants and insects were dealt with in a much more sustainable and natural way using one natural process to combat any temporary imbalance. Natural herbicides are made from peppermint castile soap, garlic, cayenne, oil of rosemary, oregano and thyme. Fungicide ingredients include Neem Oil (Azadirachta indica). Integrated pest management utilizes beneficial insects, such lace wings, which prey upon insect pests, such as aphids. Use corn gluten as a natural pre-emergent weed control. And mulch plants to suppress emerging weeds.

Provide a Removable Doormat

Providing a doormat will remove many particulates that might otherwise be tracked inside. Clean your doormat frequently.

Consider "Living Walls"

One of the newest methods of improving indoor air quality is living walls (See "Green Walls" above). Plants grow vertically in a designed structure with a hydroponic growing medium. Dirty air goes through plant roots before returning to the ducting. Roots trap and benefit from the nutrient value of fine particles and VOCs. Consequently, the return air is cleaner.

"Living walls" have the obvious benefits of attracting attention and an aesthetic benefit. Typically, people would rather look at a vertical green tapestry than concrete; in this way, it may even have a psychological benefit. The environmental benefits include adding organic life to a sterile environment, improved air quality and somewhat reduced polluted run-off.

So should people give up having a verdant swath of grass where they and their pets play? Not necessarily. Growing a healthy, strong, beautiful organic lawn requires not just a change in fertilizers but also a complete change in mindset. "With an organic lawn, (you're) not simply putting down fertilizers four times a year; (you're) initiating cultural practices to nurture life in the soil, and in turn, the soil sustains the grass." Transitioning to an organic lawn takes an initial investment of time, effort, and money, because there is a need to restore the lawn's soil biology and the health of the grass. But in the long run, people will save money and effort as grass grows healthy, strong and fights off pests and weeds without human intervention.

Steps to an organic, chemical-free lawn include:

1.   Thicken the lawn using grass seed of companion plants, such as clover that makes nitrogen bioavailable.

2.   Feed with compost to increase organic matter and microbial life of the soil.

3.   Water wisely in the morning to prevent fungal disease and minimize evaporation.

4.   Using a sharp blade cut grass high to suppress crabgrass. Longer grass grows deeper roots, naturally inhibits weeds and needs less irrigation.

5.   Leave clippings to decompose and add organic matter and nitrogen.

6.   Feed responsibly with organic fertilizer.

Improve Water Quality

Chlorine is a carcinogen, and yet all traditional swimming pools depend upon the use of hundreds of gallons of chlorine per year, resulting in swimmers with itchy skin, brittle hair and blood-shot eyes from the toxic chlorine soup. In fact, prevailing thought has been to render the water dead in order to make it safe. While some argue for salt water pools, they too use chlorine, NaCl (sodium chloride).

Natural swimming pools (biofiltration with plant materials) clean water. Twenty percent of the new pool construction in Europe uses biofiltration resulting in greater user satisfaction, reduced chemical costs, and living water of which we are a part. Rather than inoculating ourselves against every other, will there be a day when human beings peacefully coexist with nature? Not so long ago we did.

In the United States, today, everything comes packaged, sterilized, pasteurized and homogenized. People want that "new car smell," not knowing (or caring) that the VOCs in the "new car smell" are the chemicals that give us headaches and slowly poison us with compounds we genetically hand down to the next generation and the next... We must remember we are but a part of nature's web of life. Everything we do affects everything else, down to the smallest microbe; and in turn, what those smallest of unseen creatures do has a great bearing on our lives, indeed our very survival. For the most part the earth is a closed system, each living thing having an important role in the greater scheme of things. That is to say, the chemicals we pour on our grass to green it up, poisons the swallows, which no longer produce the numbers sufficient to eat the mosquitoes which carry West Nile Virus which kill human beings with increasing regularity. Until, we as a species realize this, we hinder the survival of species including our own.

References

1.   John Trotti, "Saving the planet… Huckster Style" *Erosion Control* magazine, March/ April 2008: 101.
2.   According to a recent study by the California Energy Commission (CEC)
3.   U.N. State of the World's Forests report (http://www.fao.org/docrep/009/a0773e/ a0773e00.HTM).
4.   *E* magazine January/February 2008, pg 12
5.   Proceedings of the National Academy of Science (June 12, 2007) (http://www.uore-gon.edu/newsstory.php?a=6.6.07-Crops-Jen.Fox.html)
6.   http://water.usgs.gov/nawqa/
7.   Why Is Cancer Killing Our Pets? By Deborah Shaw (Healing Arts Press, 2000)
8.   Reprinted with permission of *Organic Gardening* magazine.

Chapter 8

Recycling Guide

EDITOR'S COMMENTS

Rather than print out this chapter (which would have been updated by the time this printed), below is a link to the information.

The Santa Barbara Recycling Guide (http://www.lessissmore.org) is a great example of how to set up a recycling program.

There are many other great links available via your local communities and this is important because different communities recycle different types of materials (for example: #1 plastic versus #5 plastic).

Chapter 9

Green Cleaning

Mary J. Conrad
President, Conrad's Cleaning Company, Vero Beach, Florida

Deborah A. Pate
Vice President, Conrad's Cleaning Company, Vero Beach, Florida

WHY CLEAN "GREEN?"

There are several reasons why you should want your facility to go green, including: improved health of your facilities occupants, pollution prevention, utility savings, increased worker productivity, improved employee retention and recruitment, as well as positive marketing/public relations for your facility. But green janitorial processes are also very profitable. For example, using less toxic chemicals (which reduces material costs) improves the indoor air quality (requiring less air treatment). When a facility has good indoor air quality, the number of health-related problems of their buildings occupants goes down notably. The number of illnesses or workman compensation claims that a business has yearly definitely affects a business's bottom line. Many illnesses are as a result of allergies, sinus problems, respiratory problems, skin irritations just to name a few. This is important for employees and visitors/clients.

GREEN CLEANING TECHNIQUES

Dust inside a facility can contribute to poor indoor air quality. Cleaning personnel must use equipment that will eliminate the dust problem. Traditional ways of cleaning to remove dust consisted of a feather duster and dusting rag with some furniture polish. These traditional ways only scattered the dust back into the air. Little dust is actually picked up using traditional dusting methods. A cleaning company, janitorial service, custodial service, etc. that uses green products and procedures will use a more effective procedure when dusting a facility to limit the amount of dust particles that will be scattered back into the indoor air.

One procedure is to use tailpack or backpack vacuums (some can trap as much as 99.9% of the dust, pet dandruff, and other allergens) that have dusting attachments in addition to using the Smart Color Cleaning System microfiber clothes. Why not suck up the dust mites, bacteria, floating particles, etc? By using a certified CRI Green Label and Green Label Plus vacuum most of the dust particles will be trapped inside the filtration system. There are several CRI (Carpet and Rug Institute) certified Green Label and Green Label Plus vacuum cleaners on the market that green cleaning companies can choose from. The key is to have a vacuum that has been certified by CRI Green Label or Green Label Plus. "CRI applies science to make carpet better and last longer. CRI works with its member companies, allies in the field, independent testing laboratories and government agencies to continually improve its best practices in product improvement, environmental responsibility and customer care. The result of this pursuit of excellence means that CRI exceeds industry standards and meets customer expectations."[1] "The Green Label and Green Label Plus testing programs, overseen by independent labs, are designed for architects, builders, specifiers and facility managers who want assurances that carpet and adhesive products meet the most stringent criteria for low chemical emissions and help improve indoor air quality. Currently, carpet, cushion and adhesives as well as vacuum cleaners are tested in these programs."[2]

Professionals using green cleaning practices also can reduce cross-contamination from one area of a facility to another by using designated equipment. For example, who wants the same equipment that was just used in the bathrooms to be used in common office spaces or the break/lunchroom areas? A color-coded system can be set up for equipment: one color to be used in certain areas. When finished in that area the equipment is changed to another color and is used in the next area to be cleaned. Using the color system harmful bacteria is not carried through the facility. Equipment that is used in green cleaning can also improve the "downtime" a janitorial service needs to have an area blocked off due to cleaning. New and improved floor cleaning systems are quieter, use less water, less cleaning product, and do not use large amounts of hot water. The equipment is ergonomically designed to help prevent injury of the person using the equipment. These are just a few methods and products that a green cleaning service can do to improve indoor air quality.

CASE STUDIES

Are you still not convinced that green cleaning improves indoor air quality? Look at the following case study numbers to understand how green cleaning really works to improve indoor air quality. Charles Young Elementary School started a green cleaning and indoor air quality program that improved school attendance, math scores and reading scores in their students. After the programs were started the school attendance increased from 89% to 93%. The student's math scores increased from 51% to 76%, while reading scores increased from 59% to 75%. The study also showed that the risk of adverse health effect is small.[3]

A second case study was conducted at Frank Porter Graham Child Development Center, Chapel Hill, NC. It was found that after the school started a green cleaning and new indoor air quality programs the amount of airborne dust inside the building decreased by 52%, the VOC concentrations decreased by 49%, bacteria decreased by 40%, and fungi colony-forming units decreased by 61%.[4]

Schools are not the only facilities that have problems with indoor air quality. Facilities that have been built without consideration for sustainability use traditional carpets, paints, flooring, ventilation systems, ceiling tiles, lighting, windows, etc. These items contribute to "sick building syndrome." Harmful VOCs are released from items such as carpeting, paints, flooring, soil, pesticides, etc. Let me that a second to explain that "Greening a facility" is more than using green cleaning methodologies and products.

GREEN PLANNING

Planning to build a green facility? Do not forget the many different trades people that will come together to get the job done. All trades people need to be on the same page. The different trades should be doing their part to make sure the facility is compliant with green standards. If remodeling a current facility to be green, seek the help of professionals that can help you get your facility green, even "certified." Facilities that are sustainable and green are healthy and provide good indoor air quality for its occupants.

Positive marketing/public relations are always good for a facility. Governments on the local, state, and national levels have encouraged businesses and government to change the way they do business. Some local level governments made their cities green through waste recycling programs, water management, utilities programs, schools, hospitals, hotels, and new construction projects (residential buildings, apartments, commercial businesses). Some state governments have encouraged employees to recycle and come up with "green" alternatives for conducting business.

Today, the green movement is not just "top-down" through an organization. Issues such as global warming, pollution of our waterways, burning of the Amazon Rain Forest, severity of hurricanes in the past few years, and melting glaciers have been in the media a lot within the past few years. People seem to be changing their way of thinking and living to be greener. Today people are looking for ways to keep their family healthier, and safer. They want to ensure their children have natural resources available to them. People seem to care more and more about their environment indoors and outdoors. The media bombards us with information about how harmful products we have lived with for years are making us sick.

The public has become green conscious in today's market. The demand for safer schools, hospital, hotels, homes, workplaces, etc. is on the rise. Let's look at Austin, Texas as an example of how local and state government is encouraging builders and contractors to build green. Austin's focus is to build sustainable housing. "Austin Energy Green Building Program has been rating homes for energy efficiency since the 1990s. In October 2007, Austin officials pledged to make all new single family homes in the city zero energy by 2015. Austin Mayor Will Wynn states he wants to have affordable housing that will lower utility bills, help improve air quality and fight against global warming. "[5] "On September 21, 2006 Virginia Governor Timothy M. Kaine help launch an initiative that provides homes that save money on energy and utility bills, provide improved indoor air quality, are durable, and feature environmentally friendly products."[6]

FACTOID: A team of nine realtors in Phoenix Arizona have gained $20 million dollars in business with people who want green facilities.[7] Facilities that are built green are valuable. The public values good healthy environments. October 2007, Professor Norman Miller of the University of San Diego conducted a study of office buildings. "Miller's study revealed that ENERGY STAR-rated office buildings (those in the top 25 percent of energy performance) since 2004 have had 2 percent greater occupancy and a $2 per square foot greater rents. To top that, in 2006, ENERGY STAR buildings sold at a 30 percent premium (in dollars per square foot) to non-ENERGY STAR-rated buildings."[8] Green facilities are good for everyone involved and help protect our environment.

A facility that has been certified green by a reputable independent entity can use that certification rating as a marketing tool. A facility can earn LEED certification by following certain USGBC guidelines. A building that is LEED certified can be marketed as such and the facility can use that as a positive marketing point to potential tenants. USGBC's LEED certification is a well known independent entity that certifies facilities as being green. Many states have their own state level green certification programs. If your facility is certified it gives it creditability take the public can trust. A green facility has increased value in the market place. People want safe, healthy environmentally friendly facilities.

Increased worker productivity can be achieved by increased ventilation and temperature control. In green facilities with green HVAC systems have less pollution blowing around from office to office. Green certified facilities have less employee absenteeism. Employees are comfortable in their workplace and want to be in that facility. The amount of work output from the workers goes up; thus, increasing the companies' bottom line.

Case Study: A corporation moved 500 people into a new 150,000 ft^2 building in Wisconsin. In their new building, most of the employee workstations were relocated to the perimeter of the building to capitalize from the use of natural light. When compared to their old building, productivity increased 16%.[9] Carnegie Mellon University studies showed productivity increases can be shown ranging from 0.4% to 18%.[10] An increase of 1% in productivity is equivalent to 5 minutes per working day, which is equal to $600-$700 per employee per year, or $3.00/ft squared per year.[11] This also helps with employee retention and recruitment.

Thus far we have looked at green buildings for what they can do for us. Now look at what green buildings can do for our environment. "Traditional buildings usually account for 12% of the potable water use, 30% of greenhouse gas emissions, 36% total energy use, 65% of waste output and 70% of electricity consumption. Heating, cooling, and electrical used by traditional U.S. buildings generate 38% of all U.S. CO_2 emissions, and 10% of the world's total CO_2 annually."[12] Green buildings are designed to use less of our natural resources. Green buildings are less taxing on our natural resources even when they have been completed and occupied. The different trades that come together to design and build a facility all contribute by using green environmentally friendly products. A facility can earn LEED certification by following certain USGBC guidelines. A facility/building that has LEED certification has been designed and built under close supervision to ensure all aspects of the building process meet its environmental guidelines. A facility that is LEED certified can assure the public and potential occupants that it is environmentally-friendly.

We have already explained how a green facility can save money in the long run. Why would a facility go to the trouble of being green and certified and not continue to keep up the cleanliness of the facility during day to day operations? Often facilities managers cut expenses by cutting down on building cleaning, yet the savings produced by the green building will more than provide funds for cleaning services.

Cleaning services and building service contractors can offer a facility points that can be use towards LEED certification. Cleaning has changed through time for the better. Just as the trades have used science to improve their products to reduce their carbon footprint within buildings so to has the cleaning industry. Cleaning professionals have used science to improve there products and methodologies to have less of an impact on health and the environment. Remember not all buildings/facilities are green yet; nor are all cleaning services and building service contractors are green either. Do your homework before you employ a cleaning company that may do business using the traditional toxic products and methodologies that are not efficient or safe.

So now you're convinced and you want your facility to go green. If you are an employee of a company occupying a facility that is not green you must first do some research to find ways your employer can make changes to be green. Find others within the facility that believe as you do. Do research so you can understand what it takes for a building to be certified green. By doing research you will be prepared to present your idea to members in the company that came make the dream of going green come true.

STEPS TO MAKE YOUR CURRENT FACILITY GREEN:

1. Find others within your department and other departments within the company that believe going green will benefit all. These people can become the team once your employer or facility manager decides to implement a green program. The team will be important for gathering information from each department and working with building managers, vendors, other tenants, worker unions, LEED

project manager, marketing reps, HR person, security, maintenance, etc. Members of the team can later help management write a green operating policy and procedures manual.

2. If your employer or facility manager is not in a green building, "plant the seed" to make the change. Gather information to present to your employer and facility manager as to how going green can be a good thing. Present your case to management.

3. Once your employer or facility manager decides to go green, implement an agreement with a common goal that will accomplish the "green facility dream."

4. Gather information within the company using surveys and analyze the information. Surveys can be useful in finding out indoor environmental problems that may not have been commonly known. Surveys and questionnaires can provide information as to ideas to correct a problem. The ideas can be gathered and presented to teams members who will have expertise.

5. Develop a plan with the team members before any renovation a plan should be developed. The plan will have completion time lines for various areas of renovation within the facility. It will also have contingency plans for operation of the facility without disrupting business.

6. Discuss products and equipment that can be used that are environmentally-friendly. If a facility goes green it must be on all levels. Environmentally-friendly practices such as recycling should be promoted. Office products, paper products, folders, carpeting, furniture, office equipment, facility specific equipment (sewing machines, printing machines, cutting equipment, etc. specific to that business) should be viewed to see if it can be adapted to be environmentally friendly. Encourage the different floors or departments to come up with new and innovative ideas to be environmentally friendly.

7. Develop procedures that will accomplish the goal. Remember special accommodations that may be needed when working with

people with special needs. During the renovations dust and allergens may cause discomfort to persons with respiratory infections. The noise level during the renovations will have to be address. The HVAC system will need to be addressed.

8. Once the facility is green, training and education should occur. Occupants, facility employees need to know what to do when they have spills, encounter water leaks, foul smells, etc. There must be a contact person that addresses each problem. In other words if a plumber is needed who will call them. If a spill happens who will call the building service contractor (janitorial staff). Training can extend into the public to educate them, thus potentially creating positive public relations experience for the business.

9. Communications will be a continued part of maintaining a green facility. Communications can be weekly or monthly e-mails with information about environmentally friendly procedures. A day can be set aside to be facility green day. Banners, signs, announcements can continue to educate occupants and challenge them to think outside of the box to make business operations more efficient. Challenge each department to set aside items not being use (folders, notebooks, baskets, furniture, etc.) in a space to others form other departments may re-claim them for their use.

10. Periodically get feedback to ensure all departments are operating efficiently and green. Get feedback from the public.

Footnotes

1. CRI website CRI Signature Programs page, CRI has it covered; http://www.carpet-rug.org/about-cri/cri-signature-programs.cfm.
2. CRI website CRI Signature Programs page, Green Label and Green Label Plus: Higher Standards for Indoor Air Quality; http://www.carpet-rug.org/about-cri/cri-signature-programs. cfm.
3. Healthy School Environment and Enhanced Educational Performance-The Case of Charles Young Elementary School, Washington, DC, Carpet and Rug Institute, January 2002.
4. Indoor Environment Characterization of a Non-Problem Building: Assessment of Cleaning Effectiveness, US Environmental Protection Agency, March 1994. (Conducted at the Frank Porter Graham Child Development Center, Chapel Hill, NC).
5. Green Builder: Defining Green Magazine, November 2007 issue, Article Seeking Sustainability Austin pushes the building envelope. P. 11.
6. www.hbav.com/EarthCraft_PR_2006.doc, Home Builders Announce Statewide "Green Building" Program for Single-Family Residences.

7. NAR: Government Affairs: Smart Growth: Realtors taking the Lead on Green Building. www.realtor.org/smart_growth.nsf/pages/greenrealtor?opendocument.
8. The Green Building Revolution Accelerates, Environmental Design+ Construction. www.edcmagcom/ArticlesColumn/BPN_GUID_9-5-2006_A_10000000000000243338.
9. Sustainable Building Case Study 06. Sustainability High Performance Buildings deliver Productivity Improvements. Seattle.Gov www.cityofseattle.net
10. BuildingGreen.Com Feature from Environmental Building News, April 2005. Making the Case for Green Buildings-Improved Worker Productivity.
11. Green Building Costs and Financial Benefits by Gregory H. Kats. Published for Massachusetts Technology Collaborative. Copyright 2003. www.cap-e.com/ewebeditpro/items/O59F3481.pdf.
12. Askin Group Power Point-The Green Monsterbiorenewablesyracuse 0307.ppt.

PART II

Getting Projects Implemented

Chapter 10

Financing Projects and Avoiding Delays

Neil Zobler, Catalyst Financial Group, Inc.
Caterina Hatcher, U.S. EPAgency, ENERGY STAR Program

EDITOR'S COMMENTS:
"Lack of money to fund a project" is a common barrier. This chapter describes financing solutions that can help good projects get implemented and avoid delay (as well as delayed savings). In short, the cost of delay may be greater than the cost to finance... so in this scenario, (which is quite common) delays are literally the equivalent of throwing money out the window.

FINANCING EFFICIENCY PROJECTS
TODAY WITH FUTURE ENERGY SAVINGS

Administrators or managers often think they must postpone the implementation of energy efficiency upgrades because they do not have the funds in their current budgets. Other barriers may include the lack of time, personnel, or expertise. As you will see later in this chapter, most organizations have access to more financial resources than they may think, and resolving the financial barrier frequently provides solutions to all the other barriers.

Postponing energy efficiency upgrades for as little as one year can prove to be an expensive decision. The U.S. Environmental Protection Agency (EPA) estimates that as much as 30% of the energy consumed in buildings may be used unnecessarily or inefficiently. The money lost due to these inefficiencies in just one year frequently totals more than all the costs of financing energy upgrades over the course of the entire financing period!

Consider this business logic: The energy efficiency project you'd like to do is not in your current capital budget; but if the costs of financing the project are less than the operating budget dollars saved from reduced utility bills, why not finance the project? The benefits of doing the project sooner rather than later are numerous, starting with improved cash flow, better facilities, using the existing capital budget for other projects, helping make your facility "green," and more. Using this logic, financing energy efficiency projects is a good business decision. Delaying the project is, however, a conscious decision to continue paying the utility companies for the energy waste rather than investing these dollars in improving your facilities.

Energy efficiency projects are unlike most other projects. With properly structured financing, you may be able to implement energy efficiency projects without exceeding your existing operating or capital budgets. Thousands of companies that participate in ENERGY STAR® know from experience that today's energy efficiency technologies and practices have saved them operating budget dollars. In fact, some of the more conservative lenders like Bank of America and CitiBank have been providing funds for energy efficiency projects for over a decade. And implementing energy efficiency projects will have a positive impact on your organization's overall financial performance as well as the environment. So why wait? This chapter will help you understand how to leverage your savings opportunities, which financing vehicles to consider, and where to find the money.

ENERGY EFFICIENCY AND GREEN BUILDINGS

Energy efficiency and indoor air quality are critical components of any green building project and are usually required elements in green building certification programs. For instance, the U.S. Green Building Council's (USGBC) certification for existing buildings, LEED-EB, requires that certain energy efficiency prerequisites be met before a certification at any level can be achieved. LEED-EB requires that buildings achieve specific energy performance targets on EPA's energy performance rating system to meet the prerequisites (see the FINANCING TOOLS AND RESOURCES from ENERGY STAR Section below for more information on EPA's energy performance rating system). Earning the additional points above the prerequisites in LEED-EB equal to the ENERGY STAR on EPA's energy performance rating system can demonstrate that a building performs among the most energy efficient buildings in the nation.

Energy efficiency related building improvements are synergistic with other categories typically found in green building certification processes (Sustainable Sites, Water Efficiency, Energy and Atmosphere, Materials and Resources, and Indoor Environmental Quality).

Using the operating budget savings realized from implementing energy efficiency projects is a good way to offset the costs of making your building greener and obtaining LEED certification.

OPERATING VERSUS CAPITAL BUDGETS

Before addressing different financing options and vehicles, let's review some "accounting 101" fundamentals. Organizations make purchases by spending their own cash or by borrowing the needed funds. The impact on the balance sheet is either exchanging one asset for another when spending cash, or adding to the assets and liabilities when incurring debt.

To argue the advantages of one financing option versus another, it is important to be conversant with the roles of the operating expense budget and the capital expense budget in your organization. Capital expenses are those that pay for long-term debt and fixed assets (such as buildings, furniture, and school buses) and whose repayment typically extends beyond one operating period (one operating period usually being 12 months). In contrast, operating expenses are those general and operating expenses (such as salaries or supply bills) incurred during one operating period (again, typically 12 months). For example, repayment of a bond issue is considered a capital expense, whereas paying monthly utility bills is considered an operating expense.

The disadvantages associated with trying to use capital expense budget dollars for your energy efficiency projects are as follows: (1) current fiscal year capital dollars are usually already committed to other projects; (2) capital dollars are often scarce, so your efficiency projects are competing with other priorities; and (3) the approval process for requesting new capital dollars is time consuming, expensive, and often cumbersome.

When arranging financing for energy efficiency projects in the private sector, one of the most frequently asked questions is, "How do we keep this financing off our balance sheet?" and thereby not reflect the transaction on the company's financial statements as a liability or debt. The reasons for this request vary by organization and include: (a) treating the repayment of the obligation as an operating expense thereby avoiding the entire capital budget process, and, (b) avoiding the need for compliance with restrictive covenants that are frequently imposed by existing lenders, which may be viewed as cumbersome to point of interfering with the ongoing management of the company. Restrictive covenants start by requiring the borrower to periodically provide financial statements that enable the lender to track the performance of the company by calculating key financial ratios measuring liquidity (i.e., the current ratio, which is current assets versus current liabilities), leverage (debt-to-equity), and profitability margins. Covenants include maintaining financial ratios at agreed standards. If the ratios are not in compliance with these targets, the lender can call in all loans, creating serious cash flow problems for the borrower. Many of these financial ratios are improved by keeping debt off the balance sheet. Other typical covenants include limitations on issuing new debt, paying dividends to stockholders, and selling assets of the company.

While organizations in the public sector may not have to deal with the profitability and equity issues of the private sector, they do face their own challenges when incurring debt through the capital budget process, which is established by statute, constitution, or charter, and usually requires voter approval. Public sector organizations may find that the political consequences of incurring new debt may be more of a deterrent than the financial ones, particularly when raising taxes is involved.

Treating repayment of the financing for energy efficiency projects as an operating expense can keep the financing "off balance sheet." And, because the immediate benefit of installing energy efficiency projects is reducing the operating expense budget earmarked for paying the energy and water bills, off balance sheet treatment makes sense. Post-ENRON, however, having your auditors treat financings as "off balance sheet" is becoming increasingly difficult, especially in light of The Sarbanes-Oxley Act of 2002, which established new or enhanced standards for all U.S. public company boards, management, and public accounting firms. Nevertheless, several financing vehicles do allow financing payments for energy efficiency upgrades to be treated as operating expenses (e.g., operating leases, power purchase agreements, and tax-exempt lease-purchase agreements; see the FINANCIAL INSTRUMENTS section below for more information). Regardless of the type of financing vehicle used, implementing energy efficiency projects is in the best financial interests of your organization and helps protect the environment.

FINANCING TOOLS AND RESOURCES FROM ENERGY STAR

EPA's ENERGY STAR program offers a proven strategy for superior energy management with tools and resources to help each step of the way. Based on the successful practices of ENERGY STAR partners, EPA's "Guidelines for Energy Management" (available at www.energystar.gov/guidelines) illustrates how organizations can improve operations and maintenance strategies to reduce energy use and maintain the cost savings that can be realized by financing energy efficiency projects. EPA has sponsored hundreds of presentations (in person and on the Internet) on ENERGY STAR tools, resources, and best practices for organizations struggling with the challenge of making their buildings more energy efficient. One of the most common statements from participants (especially those in the public sector) is, "We don't have the money needed to do the facility upgrades; in fact, we don't even have enough money to pay for the energy audits needed to determine the size of the savings opportunity." This sentiment is simply not true because the needed funds are currently sitting in their utility operating budgets and being doled out every month to the local utilities. Organizations merely require a way to capture and redirect these "wasted energy" funds to pay for the energy efficiency projects, which will in turn create real savings. For some readers, this may seem to be "circular logic," or what may be called a "Catch 22."

Fortunately, EPA has created a number of tools and resources that, when properly used, will allow you to "break the circle" and find a path toward the timely implementation of energy efficiency projects. This section focuses on the tools that tie directly into financing such projects. These include the Guidelines to Energy Management, Portfolio Manager, Target Finder, Financial Value Calculator, Building Upgrade Value Calculator, and the Cash Flow Opportunity Calculator. All of these tools are in the public domain and available at www.energystar.gov.

Portfolio Manager

Peter Drucker's famous maxim is "If you don't measure it, you can't manage it." If your organization wants to use future energy savings to pay for the implementation of energy efficiency projects now, you must start by establishing the baseline of your current energy usage. The ENERGY STAR Portfolio Manager can help you do that. It is an interactive energy management tool that allows you to track and assess energy and water consumption across your entire portfolio of buildings in a secure online environment. Portfolio Manager can help you identify under-performing buildings, set investment priorities, verify effectiveness of efficiency improvements, and receive EPA recognition for superior energy performance.

Any building manager or owner can efficiently track and manage resources through Portfolio Manager. The tool allows you to streamline your portfolio's energy and water data, as well as track key consumption, performance, and cost information portfolio-wide. For example, you can:

- Track multiple energy and water meters for each facility
- Customize meter names and key information
- Benchmark your facilities relative to their past performance
- View percent improvement in weather-normalized source energy
- Monitor energy and water costs
- Share your building data with others inside or outside your organization
- Enter operating characteristics, tailored to each space use category within a building.

For many types of facilities, you can rate energy performance on a scale of 1–100 relative to similar buildings nationwide. Your building is not compared to the other buildings in Portfolio Manager to determine your ENERGY STAR rating. Instead, statistically representative models are used to compare your building against similar buildings from a national survey conducted by the U.S. Department of Energy's (DOE's) Energy Information Administration known as the Commercial Building Energy Consumption Survey (CBECS). Conducted every 4 years, CBECS gathers data on building characteristics and energy use from thousands of buildings across the United States. In Portfolio Manager, a building is compared to buildings in the CBECS survey that have similar building and operating characteristics. An EPA rating of 50 indicates that the building, from an energy consumption standpoint, performs better than 50% of all similar buildings nationwide, while a rating of 75 indicates that the building performs better than 75% of all similar buildings nationwide.

EPA's energy performance rating system, based on source energy, accounts for the impact of weather variations, as well as changes in key physical and operating characteristics of each building. Buildings rating 75 or higher may qualify for the ENERGY STAR label.

Portfolio Manager provides a platform to track energy and water use trends compared against the costs of these resources. This is a valuable tool for understanding the relative costs associated with a given level of performance, helping you evaluate investment opportunities for a particular building, and identifying the best opportunities across your portfolio. It also allows you to track your properties' performance from year to year.

The built-in financial tool within Portfolio Manager helps you compare cost savings across buildings in your portfolio while calculating cost savings for a specific project. Being able to quickly and clearly obtain data showing cumulative investments in facility upgrades or annual energy costs eases the decisionmaking process for best practice management of your buildings nationwide.

From a lender's perspective, a facility with a low rating is more likely to obtain larger energy savings (having more room for improvement) than a facility with a high rating. This becomes important if the energy savings are considered a primary "source of repayment" when financing energy upgrades. Portfolio Manager is also an important tracking mechanism that helps insure that the facilities are being properly maintained and the energy savings are continuing to accrue. As lenders perform due diligence on energy efficiency projects, they will become more aware of the value of Portfolio Manager.

Target Finder

Target Finder is another interactive energy management tool that helps design teams establish an energy performance target for new projects and major building renovations early in the design process. This target or goal serves as a reference for comparing energy strategies and choosing the best technologies and practices for achieving the performance goal. As the design nears completion, architects enter the project's estimated energy consumption, and Target Finder generates an energy performance rating based on the same underlying database applied to existing buildings in Portfolio Manager.

The energy use intensity generated by Target Finder reflects the distribution of energy performance in commercial buildings derived from CBECS data. The required data inputs (square footage, hours of operation, etc.) were found to be the primary drivers of energy use. The project's ZIP code is used to determine the climate conditions that the building would experience in a normal year (based on a 30-year climate average). The total annual energy use intensity for the target is based on the energy fuel mix typical in the region specified by the ZIP code. Target Finder displays the percent electricity and natural gas defaults used to calculate design targets. Users may enter their own fuel mix; but electricity must be selected as one of the choices. Site and source energy calculations are shown for both energy use intensity and total annual energy. The EPA rating, however, is calculated from source energy use.

EPA's Target Finder helps architects and building owners set aggressive, realistic energy targets and rate a design's intended energy use. Making energy efficiency a design requirement is the most cost-effective way to insure superior energy performance in your properties.

Building Upgrade Value Calculator

The Building Upgrade Value Calculator is built on a Microsoft EX-CEL™ platform and is a product of the partnership between EPA's ENERGY STAR, BOMA International, and the BOMA Foundation. The Building Upgrade Value Calculator estimates the financial impact of proposed investments in energy efficiency in office properties. The calculations are based on data input by the user, representing scenarios and conditions present at their properties. Required inputs are limited to general characteristics of the building, plus information on the proposed investments in energy efficiency upgrades.

The calculator's analysis includes the following information:

- Net investment
- Reduction in operating expense
- Energy savings
- Return on investment (ROI)
- Internal rate of return (IRR)
- Net present value (NPV)
- Net operating income (NOI)
- Impact on asset value

In addition to the above outputs, the calculator also estimates the impact the proposed energy efficiency changes will have on a property's ENERGY STAR rating.

The tool provides two ways to use its calculations: users can save and print a summary of their results, or they can generate a letter highlighting the financial value for use as part of a capital investment proposal. Because energy efficiency projects generally improve net operating income, which is an important consideration when buying and selling commercial properties, the Financial Value Calculator provides strong financial arguments in favor of implementing these projects.

Financial Value Calculator

Investments in energy performance can have a favorable impact on profit margins, earnings per share, and ultimately, shareholder value. The Financial Value Calculator, another tool built on an EXCEL™ platform, presents energy investment opportunities using the key financial metrics managers need to convey the message of improved energy performance to customers.

Cash Flow Opportunity Calculator

The Cash Flow Opportunity Calculator ("CFO Calculator") has proven to be a very effective tool, especially for public sector projects. This set of spreadsheets helps create a sense of urgency about implementing energy efficiency projects by quantifying the costs of delaying the project implementation. It was developed to help decisionmakers address three critical questions about energy efficiency investments:

1. How much of the new energy efficiency project can be paid for using the anticipated savings?
2. Should this project be financed now, or is it better to wait and use cash from a future budget?
3. Is money being lost by waiting for a lower interest rate?

Using graphs and tables, the CFO Calculator is written so that managers who are not financial specialists can use it to make informed decisions, yet it is sophisticated enough to satisfy financial decisionmakers. This tool works well for projects in both the public and private sectors.

To determine how much of the new project can be paid for using your anticipated savings, the CFO Calculator takes a practical look at your energy efficiency situation and financing opportunities. You can choose to enter either (a) best estimates of how your building currently operates and how much better it could operate, or (b) data generated when you use EPA's energy performance rating system within Portfolio Manager. Either way, the CFO Calculator provides answers to some critical financial questions in just minutes.

The first step in the process is to estimate the amount of the savings that can be captured from the existing utility budget. The working assumption is that these savings will be used to cover the financing costs and that the savings will recur. The savings amount is entered into a "reverse financial calculator," which then asks for an estimated borrowing interest rate, financing term, and the percentage of the savings you wish to use. It then calculates the amount of project improvements that could be purchased by redirecting these energy net savings to pay for the upgrades. Most organizations are surprised to learn how much new equipment and related services are "buried" in their utility bills, all of which could be installed within their existing operating budget and without spending their limited capital budget. The related services often include the initial energy audits that many feel they cannot afford, but are necessary to quantify the savings opportunity. When future energy savings are the main source of the project's repayment, the CFO Calculator becomes an effective sensitivity analysis tool that takes into account the impact of lower interest rates, longer financing terms, and utilization of savings when structuring the project's financing.

A while back, the "see how much money you are leaving on the table" argument was made to the CFO of a large city in the Northeast on behalf of the local electric utility. The CFO responded that the city was fiscally conservative and believed that waiting until funds were available in a future operating budget (thereby avoiding borrowing and paying interest) was in the best interests of the city. The CFO Calculator was used to map the cash flow consequences of these two decision points (financing now or waiting until a future budget) to demonstrate to the city's CFO and city council that financing now was a better financial decision than waiting for cash. In most instances, the lost energy savings incurred by waiting for one year are greater than the net present value of all the interest payments of most financings, making "do it now" the better financial decision. This is counterintuitive and surprises most decisionmakers. Today, this city supports the expeditious implementation of energy efficiency projects.

Another common argument for delay is waiting for a lower interest rate offering rather than financing at a higher rate that is available immediately. This situation may occur when waiting for funds from a future bond issue or for a low-cost specialty fund to replenish itself, versus accepting an immediately available third-party financing offering. The CFO Calculator allows you to compare two different interest rate offerings, and it will compute how long you can wait for the lower interest rate before the lower rate begins to cost more. It does this by including the forfeited energy savings into the decisionmaking process; truly, another "cost of delay."

CHOOSING THE RIGHT FINANCING

"Financing" should be thought of as a two-step process: financial instruments (or vehicles) and sources of funds. Once you decide which financing vehicle is best for your organization, the next step is to choose the best source of funds. Bear in mind that no one financing alternative is right for everyone. In the world of energy efficiency finance, one size definitely does not fit all! For the purposes of this discussion, our focus is limited to the public and private sectors, not consumer finance options.

Before you can choose the right financial vehicle, however, two other issues must be considered—tax exempt status and interest rates. In general, public sector organizations and some non-profits qualify for tax-exempt financing, while private sector organizations do not. Private sector organizations are usually driven by tax considerations and financial strategies, but public sector organizations do not pay taxes.

What's the interest rate?" is frequently the first question asked when evaluating financing options. Organizations able to issue tax-exempt obligations will benefit from lower interest rates than would be the case for regular "for profit" organizations. This is because, according to the Internal Revenue Code of 1986, the lender does not have to pay federal income tax on the interest earned from tax-exempt transactions. Due to competitive market forces, much of this saving is passed back to the borrower in the form of lower interest rates. Any U.S. state, district, or any subdivision thereof that (a) has the ability to tax its citizens; (b) has police powers; or (c) has the right of eminent domain and qualifies for tax-exempt financing. This includes public schools, state universities and community colleges, libraries, public hospitals, town halls, municipal governments, county governments—in summary, almost any organization that receives its funding from tax revenues. While not-for-profit organizations created under Section 501(c)(3) of the Internal Revenue Code and private organizations do not directly qualify as issuers of tax-exempt obligations, they may be able to have a "conduit agency"—a city, state, health, or education authority—apply for the financing on their behalf.

For private sector organizations, interest rate alone is rarely the best indicator of the "best deal." To show the importance of proper deal structuring, consider the following question: "Which is the better finance offering: 0% or 6%?" Most people say "0%" until they find out that the 0% obligation must be repaid in 6 months, while the 6% obligation has a term of 5 years, making "cash flow" the deciding factor. The question is further

complicated when the 6% obligation requires that the owner(s) person-
ally guarantee(s) the financing; however, the transaction can be done at
12% without personal guarantees. Once all of the terms and conditions are
known, it is easier to understand why many would prefer to pay a 12%
financing interest rate rather than 0%. In addition to the term and personal guarantees, the list of structuring points is broad and may include
whether the rate is fixed or variable, payment schedules, down payments,
balloon payments, balance sheet impact, tax treatment, senior or subordi-
nated debt, and whether additional collateral is required.

Your organization's legal structure, size, credit rating, time in business, sources of income, profitability or cash flow, type of project, general
market conditions, dollar size of the project, and the use of the equipment
being financed are also very important considerations when choosing the
right financing vehicle and source of funds.

FINANCIAL INSTRUMENTS

There are two basic approaches to funding projects: "pay-as-you-go"
and "pay-as-you-use." Pay-as-you-go means paying for the project out of
current revenues at the time of expenditure, in other words, paying cash.
If you don't have the cash, the project gets postponed until you do. Pay-
as-you-use means borrowing to finance the expenditure with debt service
payments being made from revenues generated during the useful life of
the project. Because energy efficiency projects generate operating savings over the life of the project, the pay-as-you-use approach makes good
sense.

As previously mentioned, public sector organizations can borrow at
tax-exempt rates, which are substantially lower than the taxable rates that
private sector organizations will pay when financing. While tax-exempt
financial instruments can only be used by public sector (and qualifying
non profit and private sector) organizations, taxable instruments can be
used by all. This section will help identify the financial instruments best
suited for public or private sector application. Public sector instruments
include bonds and tax-exempt lease purchase agreements. Private sector
instruments include a variety of commercial leases. All sectors can benefit
from energy performance contracts and power purchase agreements. Bear
in mind that there are exceptions to every rule; and structuring a financing
to comply with tax or budget issues is often complicated, which is why

working with a financial advisor may prove helpful.

Major capital projects are funded by some form of debt, which is categorized as either short term (for periods of less than one year) or long term (for periods greater than one year). Most borrowings by public sector organizations require citizen approval, either directly through referendum or indirectly through actions of an appointed board or elected council. However, revenue bonds and tax-exempt lease-purchase agreements may not require local voter approval (see details below).

Most of us are familiar with traditional loans, which are debt obligations undertaken by a borrower. The title of the asset being financed is typically in the name of the borrower and the lender files a lien on the asset being financed, prohibiting the borrower from selling the asset until the lien is lifted. Banks frequently require additional collateral, which may take the form of keeping compensating balances in an account or placing a blanket lien on all other assets of the organization. A conditional sales agreement or installment purchase agreement is a kind of loan that is secured by the asset being financed; the title of the asset transfers to the borrower after the final payment is made. All loans are considered "on balance sheet" transactions and are common to both the private and public sectors.

Frequently used short-term debt instruments include bank loans (term loans or lines of credit), anticipation notes (in anticipation of bond, tax, grant or revenues to be received), commercial paper (taxable or tax-exempt unsecured promissory note that can be refinanced or rolled over for periods exceeding one year), and floating-rate demand notes (notes that allow the purchaser to demand that the seller redeem the note when the interest rate adjusts).

Long-term debt is frequently in the form of bonds. Commercial bonds can be quite complex (asset backed, callable, convertible, debenture, fixed or floating rate, zero coupon, industrial development, etc.) and usually require working with an investment banker. In the public sector, bonds fall into two categories: general obligation (GO) bonds and revenue bonds. GO bonds are backed by the issuer's full faith and credit and can only be issued by units of government with taxing authority. Because the issuer promises to levy taxes to pay for these obligations, if necessary, these bonds have the lowest risk of default and, therefore, the lowest cost. Interest paid on GO bonds is typically exempt from federal income taxes and may be exempt from state income taxes.

Revenue bonds are also issued by local governments or public agencies. However, because they are repaid only from the specific revenues named in the bond, they are considered to be riskier than GO bonds. Revenue bonds may not require voter approval and often contain covenants intended to reduce the perceived risk. Typical covenants include rate formulas, the order of payments, establishing sinking funds, and limiting the ability to issue new debt. Small municipalities that have difficulty issuing debt often add credit enhancements to their bonds in the form of bond insurance or letters of credit.

In the case of most energy efficiency projects, the source of repayment is the actual energy savings (considered part of the operating budget) realized by the project. When the approval process to obtain the necessary debt is a barrier, public sector organizations may be able to limit the repayment of the financing costs to their operating budget by using a tax-exempt lease purchase agreement. This solution may avoid the capital budget process altogether.

Tax-exempt Lease-purchase Agreements

Tax-exempt lease-purchase agreements are the most common public sector financing alternatives that are paid from operating budget dollars rather than capital budget dollars. A tax-exempt lease purchase agreement is an effective alternative to traditional debt financing (bonds, loans, etc.) because it allows a public organization to pay for energy upgrades by using money that is already set aside in its annual utility budget. When properly structured, this type of financing makes it possible for public sector agencies to draw on dollars to be saved in future utility bills to pay for new, energy-efficient equipment and related services today.

A tax-exempt lease-purchase agreement, also known as a municipal lease, is closer in nature to an installment-purchase agreement than a rental agreement. Under most long-term rental agreements or commercial leases (such as those used in car leasing), the renter or lessee returns the asset (the car) at the end of the lease term, without building any equity in the asset being leased. In contrast, a lease-purchase agreement presumes that the public sector organization will own the assets after the term expires. Further, the interest rates are appreciably lower than those on a taxable commercial lease-purchase agreement because the interest paid is exempt from federal income tax for public sector organizations.

In most states, a tax-exempt lease-purchase agreement usually does not constitute a long-term "debt" obligation because of non-appropriation language written into the agreement and, therefore, rarely requires public approval. This language effectively limits the payment obligation to the organization's current operating budget period (typically a 12-month period). The organization will, however, have to assure lenders that the energy efficiency projects being financed are considered of essential use (i.e., essential to the operation of your organization), which minimizes the non-appropriation risk to the lender. If, for some reason, future funds are not appropriated, the equipment is returned to the lender; and the repayment obligation is terminated at the end of the current operating period without placing any obligation on future budgets.

Public sector organizations should consider using a lease-purchase agreement to pay for energy efficiency equipment and related services when the projected energy savings will be greater than the cost of the equipment (including financing), especially when a creditworthy energy service company (ESCO) guarantees the savings. If your financial decisionmakers are concerned about exceeding operating budgets, they can be assured that this will not happen because lease payments can be covered by the dollars to be saved on utility bills once the energy efficiency equipment is installed. Utility bill payments are already part of any organization's normal year-to-year operating budget. Although the financing terms for lease-purchase agreements may extend as long as 15 to 20 years, they are usually less than 12 years and are limited by the useful life of the equipment.

There may be cases, however, when tax-exempt lease-purchase financing is not advisable for public sector organizations; for example, when (1) state statute or charter may prohibit such financing mechanisms; (2) the approval process may be too difficult or politically driven; or (3) other funds are readily available (e.g. bond funding that will soon be accessible) or excess money exists in the current capital or operating budgets.

How is Debt Defined in the Public Sector?

It is important for managers to be aware of the different interpretations of "debt" from three perspectives—legal, credit rating, and accounting. As mentioned above, most tax-exempt lease-purchase agreements are not considered "legal debt," which may prevent the need to obtain voter approval in your locality. However, credit rating agencies, such as Moody's and Standard & Poor's, do include some or all of the lease-purchase obligations when they evaluate a public entity's credit rating and its ability to meet payment commitments ("debt service"). These two perspectives (legal and credit rating) may differ markedly from the way lease-purchase agreements are treated (i.e., which budget is charged) by your own accounting department and your organization's external auditors.

In general, lease-purchase payments on energy efficiency equipment are small when compared to the overall operating budget of a public organization. This usually means that the accounting treatment of such payments may be open to interpretation. Because savings occur only if the energy efficiency projects are installed, the projects' lease-purchase costs (or the financing costs for upgrades) can be paid out of the savings in the utility line item of the operating budget. Outside auditors may, however, take exception to this treatment if these payments are considered "material" from an accounting perspective. Determining when an expense is "material" is a matter of the auditor's professional judgment. While there are no strictly defined accounting thresholds, as a practical guide, an item could be considered material when it equals or is greater than 5% of the total expense budget in the public sector (or 5% of the net income for the private sector). For example, the entire energy budget for a typical medium-to-large school district is around 2% of total operating expenses; therefore, so long as the payments stay under 2%, energy efficiency improvements will rarely be considered "material" using this practical guideline.

What are Energy Performance Contracts?

Energy efficiency is big business in the United States. A recent Lawrence Berkeley National Laboratory-National Association of Energy Service Companies (LBNL-NAESCO) company survey has ESCOs projecting their 2008 revenues at more than $5 billion and growing at an annual rate of 22% since 2006.

The survey went on to confirm that "MUSH" markets—municipal and state governments, universities and colleges, K-12 schools, and hospitals—have historically represented the largest share of ESCO industry activity, which was 58% of industry revenues in 2006. The federal market represented 22% and public housing 2% of revenues, while the industrial sector represented 6% and the commercial sector 9%. Energy performance contracting plays an important role in the implementation of these projects.

In most parts of the United States, an energy performance contract (EPC) is a common way to implement energy efficiency improvements. It frequently covers financing for the needed equipment, should your organization chose not to use internal funds. In fact, every state (except Wyoming) has enacted some legislation or issued an executive directive to deal with energy efficiency improvements. While EPCs are used both in the public and private sectors, 82% of the revenues of ESCOs come from public sector clients. Properly structured EPCs can be treated as an operating, rather than a capital expense.

If you search for the phrase "energy performance contract," a variety of definitions appear. The U.S. Department of Housing and Urban Development (HUD) says that a performance contract is "an innovative financing technique that uses cost savings from reduced energy consumption to repay the cost of installing energy conservation measures..." The Energy Services Coalition—a national nonprofit organization composed of energy experts working to increase energy efficiency and building upgrades in the public sector through energy savings performance contracting—states that a performance contract is "an agreement with a private energy service company (ESCO)... [that] will identify and evaluate energy-saving opportunities and then recommend a package of improvements to be paid for through savings. The ESCO will guarantee that savings meet or exceed annual payments to cover all project costs... If savings don't materialize, the ESCO pays the difference..."

Notice that both definitions mention "payment." Let's dig a little deeper.

DOE claims that "Energy performance contracts are generally financing or operating leases provided by an energy service company (ESCO) or equipment manufacturer. What distinguishes these contracts is that they provide a guarantee on energy savings from the installed retrofit measures, and they usually also offer a range of associated design, installation, and maintenance services." The NY State Energy Research and Development Authority (NYSERDA) states that "An EPC is a method of implementation and project financing, whereby the operational savings from energy efficiency improvements is amortized over an agreed-upon repayment period through a tax-exempt lease purchase arrangement..." Meanwhile, the Oregon Department of Energy says, "An energy savings performance contract is an agreement between an energy services company (ESCO) and a building owner. The owner uses the energy cost savings to reimburse the ESCO and to pay off the loan that financed the energy conservation projects."

So, what is the funding mechanism used in an EPC? Is it a financing or operating lease (which are two very different structures, see EQUIPMENT LEASING below), a tax-exempt lease-purchase agreement, or a loan? From a financing perspective, these are all different vehicles with diverse accounting and tax consequences. The answer is yes to all.

The definition of a performance contract may be found in some state statutes; however, in general, it is not clearly defined and usually includes a variety of services such as energy audits, designing, specifying, selling and installing new equipment, providing performance guarantees, maintenance, training, measurement and verification protocols, financing, indoor air quality improvements, and more. One major benefit of using a performance contract is the ability to analyze the customer's needs and craft a custom agreement to address the organization's specific constraints due to budget, time, personnel, or lack of internal expertise. This includes choosing the financing vehicle that best suits the organization's financial and/or tax strategies.

Designed for larger projects, performance contracting allows for the use of energy savings from the operating budget (rather than the capital budget) to pay for necessary equipment and related services. Usually there is little or no upfront cost to the organization benefiting from the installed improvements, which then frees up savings from reduced utility bills that would otherwise be tied up in the operating budget. An energy performance contract is an agreement between the organization and an ESCO to provide a variety of energy saving services and products. Because these improvement projects usually cover multiple buildings and often include upgrades to the entire lighting and HVAC systems, the startup cost when not using an EPC may be high and the payback period lengthy. Under a well-crafted EPC, the ESCO will be paid based on the verifiable energy savings.

The ESCO will identify energy saving measures through an extensive energy audit, and then install and maintain the equipment and other upgrades. This includes low- and no-cost measures which contribute to the projects overall savings. The ESCO works closely with the client throughout the approval process to determine which measures to install, timing of the installations, staffing requirements, etc. The energy savings cover the costs of using the ESCO and financing for the project.

The most common type of performance contract is called a "guaranteed savings agreement," whereby the ESCO guarantees the savings of the installed energy-efficiency improvements (equipment and services). The ESCO assumes the performance risk of the energy-efficient equipment so that if the promised savings are not met, the ESCO pays the difference between promised savings and actual savings. If the savings allow, a performance contract may include related services such as the disposal of hazardous waste from the replacement of lighting systems, or from the removal of asbestos when upgrading ventilation systems. The ESCO usually maintains the system during the life of the contract and can train staff to assist or to continue its care after the expiration of the contract period. The ESCO can also play a major role in educating the customer organization about its energy use and ways to curb it.

A shared savings agreement is another type of energy services performance contract under which the ESCO installing the energy-efficient equipment receives a share of the savings during the term of the agreement. In a fixed shared savings agreement, the customer agrees to a payment based on stipulated savings, and once the project is completed, the payments usually cannot be changed. After the completion of the project, the savings are verified by an engineering analysis or other mutually agreed upon method. In a true shared savings agreement, the savings are verified on a regular basis; and the savings payments change as the savings are realized.

In summary, performance contracts typically contain three identifiable components: a project development agreement indicating which measures will be implemented to save energy (and money); an energy services agreement indicating what needs to be done after the installation to maintain ongoing savings; and a financing agreement. Organizations may choose to finance the projects independently of the ESCO, especially when they can access lower cost financing on their own (as in the case of public sector organizations when accessing tax-exempt funding). It is important to note that savings are measured in kWh and therms, and then translated into dollars at the current market price for electricity and natural gas.

Regardless of the type of energy services agreement, it is important to remind the reader of two critical components that are needed to ensure that the energy performance and operational goals are met: (1) commissioning, and (2) measurement and verification. Commissioning is the process of making sure a new building functions as intended and communicating the intended performance to the building management team. This usually occurs when the building is turned over for occupancy. Ongoing and carefully monitored measurement and verification protocols are vital to ensure the continuing performance of the improvements, especially when the energy savings are the source of the financing repayment.

Power purchase agreements (PPAs), also know as design–build–own–operate agreements, are ones in which the customer purchases the measurable output of the project (e.g., kilowatt hours, steam, hot water) from the ESCO or a special purpose entity (SPE) established for the project, rather than from the local utility. And they purchase at lower rates or on better terms than they would have received by staying with the utility. These agreements work well for on-site energy generation and/or central plant opportunities. PPAs are frequently used for renewable energy and cogeneration projects (also known as combined heat and power projects). Due to the complexities of the contracts, projects using PPAs are typically very large. PPAs are frequently considered "off balance sheet" financings and are used in both the public and private sectors.

Commercial Leasing

Energy efficiency equipment that is considered by the Internal Revenue Service (IRS) as personal property (also know as "movable property" or "chattels") may be leased. The traditional equipment lease is a contract between two parties in which one party is given the right to use another party's equipment for a periodic payment over a specified term. Basically, this is a long-term rental agreement with clearly stated purchase options that may be exercised at the end of the lease term. Commercial leasing is an effective financing vehicle and is often referred to as "creative financing." Leases can be written so the payments accommodate a customer's cash flow needs (short-, long-, or "odd-" term; increasing or decreasing payments over time; balloon payments; skip payments, etc.). Leases are frequently used as part of an organization's overall tax and financing strategy and, as such, are mostly used in the private sector.

From a financial reporting perspective, however, commercial leases fall into only two categories (an operating lease or a capital lease); each has substantially different financial consequences and accounting treatment. The monthly payments of an operating lease are usually lower than loan payments because the asset is owned by the lessor ("lender"), and the lessee's ("borrower's") payments do not build equity in the asset. The equipment is used by the lessee during the term, and the assumption is that the lessee will want to return the equipment at the end of the lease period. This means that the lease calculations must include assumptions that the residual value of the leased asset can be recovered at the end of the lease term. In other words, equipment that has little or no value at the end of the lease term will probably not qualify under an operating lease. For example, lighting systems would not qualify, while a well maintained generator in a cogeneration project might. Operating leases are considered "off balance sheet" financing, and payments are treated as an operating expense.

A common capital lease is a "finance lease," which is similar to a conditional sales agreement because the asset must be reflected on the lessee's (borrower's) balance sheet. A finance lease is easily recognized because the customer can buy the equipment at the end of the lease term at a stated price that is less than its fair market value ("bargain purchase option"). For example, a lease with a one dollar purchase option is clearly a capital lease. Other conditions that define a capital lease deal with the term of the lease, transfer of ownership, and lessor's equity in the asset.

Not everyone realizes that the tax treatment of a lease may be different from the financial reporting treatment of a lease. A tax lease or guideline lease is one in which the lessor keeps the tax incentives provided by the tax laws for investment and ownership of equipment (typically depreciation and tax credits). Generally, the lease rate on tax or guideline leases is reduced to reflect the lessor's recognition of this tax incentive. A true lease is similar to a long-term rental agreement, gives the lessee the option to buy the equipment at its true fair market value at the end of the lease term, and may allow the lessee to deduct the monthly lease payments as an operating expense for income tax purposes. After all, you can't depreciate an asset that you do not own. However, a true lease will be picked up on the balance sheet.

Public sector organizations frequently lease equipment. However, because most public sector organizations are tax-exempt, tax strategies are not usually a consideration when deciding which type of lease to enter into.

FUNDING SOURCES

Once you have determined that internal sources of funds are not available or are insufficient for your energy efficiency project, your options become (a) using third party lenders, (b) postponing the project, or(c) installing part of the upgrades by breaking the project into smaller pieces. Earlier in this chapter, we explained why postponing or delaying the project can, in fact, be the most expensive alternative. So financing becomes the best decision. Once you decide to finance the project and identify a preferred financing vehicle, the next step is to evaluate potential funding sources.

Traditional funders include banks, commercial credit companies leasing companies, insurance companies, brokerage houses, and vendors. If you are dealing with a large financial institution, it is important to contact the right department within that organization obtain the best pricing for your project. For example, when speaking to their bank, most people start with their "regular banker" who may limit their discussions to loans. Keep in mind, however, that larger banks have a public finance department where you will find tax-exempt lease purchasing, and sometimes even a bank-owned leasing company where you may structure a special equipment lease. Large commercial credit companies may divide the market by the size of the transaction—small ticket, middle market, and large ticket. To further complicate matters, companies often define their market divisions differently (i.e., under $25k for micro-ticket, $25k–$500k for small ticket, $500k–$15 million for middle market, and over $15 million for large ticket). If your project is $500,000, the small ticket and middle-market groups may quote you two different prices, even though they work for the same lender. Add the dimension of public-sector versus private-sector finance, and you may get different pricing again. This is especially true in organizations where the sales staff's compensation plan is commission based, and they do not have a company lead-sharing policy in place.

In almost every state where the electric industry has been restructured (deregulated), legislation has been passed to create a system benefits charge (also know as a public benefits charge) that adds a defined surcharge (fee) to the electricity bills. These fees are used to support energy-related projects that provide public benefits such as renewable energy, energy efficiency, low-income customer programs, energy R&D, or other related activities that may not be available in a competitive market. These fees, usually a per-kilowatt (kWh) hour cost or a fixed charge, are charged to all customers and cannot be bypassed. The amount of the charge varies by state and accumulates in a fund, which is usually administered by the state's energy office or local utility.

If you are located in a state with public benefits charges (see map below), you may find that your energy efficiency project qualifies for a low-cost or below-market financing program funded by these charges. A listing of the total funding amounts by category and administrative contact information is available at the American Council for an Energy-Efficient Economy's web site (http://www.aceee.org/briefs/aug07_01.htm). In the regulated states, the public utility commissions continue to provide incentives (rebates, loans, grants, etc.) to promote energy efficiency and renewable energy projects.

Examples of some of these programs include Texas' LoanSTAR Revolving Loan Program (available to schools, local governments, state agencies, and hospitals) which will loan up to $5 million at 3% interest (today's rate). In Oregon, the Ashland Electric Utility offers 0% loans to their commercial customers to finance energy efficiency improvements to their facilities. The loans can be used for lighting retrofits and other energy-efficiency measures.

Virtually every state offers some form of incentive, and we recommend starting your funding search by reviewing these special programs. A good place to find a list of energy efficiency incentives is the Database of State Incentives for Renewables and Efficiency (www.dsireusa.org).

COMMERCIAL PROPERTIES

Commercial properties use a lot of energy. In fact, energy is the single largest operating expense in an office building, representing about 30% of a typical building's costs.

Forward thinking members of corporate America and many owner-occupiers of commercial real estate have aggressively adopted energy efficiency measures as a way to improve their cash flows, asset values, and the environment. ENERGY STAR has hosted an annual awards ceremony since 1993 to acknowledge the efforts of these organizations. In 2008 alone, the ENERGY STAR awards honored industry and brand leaders, including USAA Real Estate Company, TIAA-CREF, CB Richard Ellis Group, Transwestern Commercial Services, Simon Property Group, J.C. Penney Company, Food Lion, Giant Eagle, and Marriott International. The contributions to their stockholders, owners, and the environment from these firms' energy efficiency efforts are real and quantifiable.

Some examples of ENERGY STAR Award winners include USAA Real Estate Company, which owns and operates more than 38 million square feet of real estate and has reduced energy consumption by approximately 28 percent since 2001—the equivalent of avoiding over 121 million pounds of carbon dioxide emissions or planting over 16,500 acres of trees. Trizec Properties launched a portfolio-wide energy conservation and management program and reduced energy consumption by approximately 15%, translating into cost savings of almost $16 million annually. Transwestern Commercial Services performed a series of energy-efficient upgrades across its portfolio in 2004, saving an average of 15 to 30 percent per building. Hines, one of the largest private real estate organizations in the world, has a portfolio that includes more than 1,000 properties representing approximately 416 million square feet. Hines' long-time philosophy is to build better, greener and more energy-efficient buildings, and to offer office space at competitive rental rates. Hines finds that its sustainability platform provides a marketing advantage with tenants and brokers who are becoming more aware of the value of "green" everyday.

However, energy efficiency retrofit projects in the commercial real estate sector (rental properties in particular) must deal with a variety of challenges that are not faced by other sectors. These challenges include:

- Split incentives (when the tenant and not the building owner pays the utility bills)
- Properties held under complex legal structures that make it difficult and time-consuming to obtain financing (limited liability companies, limited partnerships, etc.)
- Non-recourse financing already in place but new lenders want recourse on traditional energy efficiency project financings, especially when the equity in the property is highly leveraged
- Difficulty in obtaining a security interest in the new energy assets being installed unless the entire building is refinanced
- Tenants not wanting to take on debt for long-term leasehold improvements
- Lenders not wanting to provide secondary financing to building owners for terms longer than the remaining lease term for key tenants
- Properties often managed by professional, third-party management companies that generally cannot enter into debt obligations on behalf of owners without special authority
- Building owners and/or tenants unable or unwilling to borrow more money as they may be concerned about reaching their debt capacity or violating covenants in existing loan agreements

In 1991, Pacific/Utah Power, the electric generation and distribution divisions of PacifiCorp, implemented a program that addressed these challenges by offering their retail customers an innovative type of efficiency program in which customers repaid the costs of their efficiency installations through monthly energy service charges on their electric bills. As part of the utility's demand-side management program, the utility had recourse to shutting off power if the customer defaulted. In essence, the electric meter became the "credit." The utility did more than 1,000 transactions, and eventually sold its loan portfolio to a major U.S. bank.

The many benefits of this program if applied to the commercial real estate sector would include not requiring the tenant or building owner to enter into a debt obligation thereby overcoming aversions to borrowing, tying the repayment to the use of energy-saving equipment, allowing the building owner to acquire new energy saving equipment at no direct cost, and having the tenants reduce their monthly utility bills without incurring debt.

While the PacifiCorp program is no longer available, this financing model continues to offer benefits that would be appreciated by the commercial real estate sector. The challenges to making this program work today include persuading the appropriate state public utility commissions to authorize a new tariff (energy service charge), finding a lender willing to underwrite the program, and insuring the installed equipment works as promised. Perhaps a "bill-to-the-meter" financing program will reappear and be added to the types of financing options mentioned earlier in this chapter. Meanwhile, the financing choices for commercial properties are clear: pay cash, add it to the mortgage (refinance or take out a second mortgage), take out a loan, lease it, or enter into an energy performance contract. As long as the energy efficiency projects get done and the buildings are managed with energy efficiency as a top priority, everybody wins!

CONCLUSION

This chapter demonstrated how public and private sector organizations can redirect energy inefficiencies and waste from their current and future operating budgets in order to pay for the needed energy efficiency improvements today. Practical suggestions that support the urgency of implementing these projects were shared and methods to quantify the costs of delay outlined. One section reviewed useful, field-tested tools and resources that are offered through EPA's ENERGY STAR program— all in the public domain for you to use. Finally, potential sources of low- or no-cost funding for your projects were presented, along with a variety of alternative financing vehicles.

It should be clear from this chapter that a decision to put off installation of more energy-efficient equipment and implement related energy-saving measures is a decision to continue paying higher utility bills to the utility company. Using the captured energy savings to pay for the financing of improvements is recommended, essentially making them "self-liquidating" obligations.

Because energy efficiency projects pay for themselves over time, the bottom line is that the prompt implementation of properly maintained energy efficiency improvements is simply a good business decision.

Chapter 11

Get Solar on Your Roof
For Free—
The Power Purchase Agreement

Ryan Park

INTRODUCTION:

The installation of a solar electricity project is a great way to green your business and the financial return can be very attractive if your business is positioned in a state or utility district offering rebates and/or tax credits for solar installations. Due to the fact solar electricity panels will last many decades and have a significant amount of embedded energy, the upfront cost of a solar project can be significant. At this time there are several ways to proceed with a solar electricity project: cash purchase, project finance, capital or operational lease, or power purchase agreement (PPA). Due to the many benefits of a PPA, over 60 percent of all commercial solar projects are financed using this structure. In this chapter we will explain the Power Purchase Agreement and its advantages over other forms of financing. Then we will proceed into design criteria to maximize your return on investment if pursuing a PPA.

The primary benefit for using a PPA is to get a solar system on your roof with zero upfront costs, while also having a known price for energy (that the solar system produces) for 20 years- which reduces risk from energy supply price spikes.

PPA Advantages

	FINANCE LEASE	TRUE LEASE	CAPITAL PURCHASE	POWER PURCHASE AGREEMENT
Upfront Capital?	NONE	NONE	YES	NONE
Performance Risk?	YES	YES	YES	NONE
System Expertise Required?	YES	YES	YES	NONE
Maintenance Required?	YES	YES	YES	NONE
Purchase Required?	YES	YES – but option to re-lease	YES	NONE

WHAT IS A POWER PURCHASE AGREEMENT (PPA)?

A power purchase agreement or PPA is a long-term agreement to buy power from a company that produces electricity. A third-party financier will provide the capital to build, operate, and maintain a solar electricity installation for 15 to 20 years. The host customer is only responsible for purchasing the electricity produced by the solar system. It is the responsibility of the PPA provider to assume all risks and responsibilities of ownership. A PPA provider will own, operate, maintain, and clean the system for optimal performance. The PPA provider will also have sophisticated real-time monitoring services to verify the system is working properly.

The host customer will run their business as usual, without any concerns about how to best operate the solar power system. At the end of the PPA term, the system can be purchased by the host at fair market value or the PPA can be renewed. Overall, a PPA enables a host customer (and our world) to benefit from the use of "green" energy, while still receiving some of the benefits of ownership through lower electricity costs and an improved public image. It also allows the host customer to spend their capital budget on their core business instead of on a solar electricity system.

How is it Different from Other Forms of Financing?

A PPA is unique because there are zero upfront costs and the host only pays for the power produced and has no responsibility for maintaining the system. If a system were financed through project financing or a

PPA Energy Services Model
Power Purchase Agreement

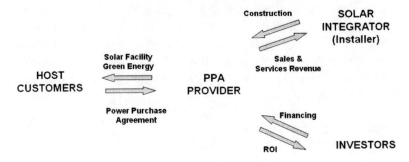

lease, the host would be required to make payments on the loan regardless of the functionality of the system. While solar electricity systems are not prone to maintenance problems, few businesses want to add solar electricity system oversight to an already overloaded list of issues to deal with for their core business.

TYPICAL PPA TERMS

Zero Up-front Cost
15- 20 year PPA Agreement
This is the term you will be required to purchase all electricity generated by the solar system. PPA providers realize that solar electricity systems will last decades. They are willing to purchase the system to own, operate, and maintain the system over 15-20 + years when traditional financing is only a third to half that time frame. The result is that many PPA agreements can be cash flow positive from the start, which means the price you will pay for solar power is actually less than you would pay for traditional electricity from the utility. Keep in mind that when entering a PPA you will still receive power from the utility company to provide electricity when solar is not able to provide all of your electricity needs; night or heavy power usage facilities.

Starting kWh Price
Depending on the location of the solar installation, size of the project, and other parameters affecting power production, a PPA will have a

specific kWh starting price.

PPA kWh Price Inflation

In order to reduce the starting kWh price as much as possible, the PPA agreement will have a set kWh price inflation per year. For virtually all PPA agreements, the contracted inflation rate is lower than the historical electricity inflation of the local utility. This allows a host customer to forecast what their power prices will be in 10-20 years and likely lead to significant savings over time.

Comparative Cost of Electricity:
Utility vs. Solar Over Life of System

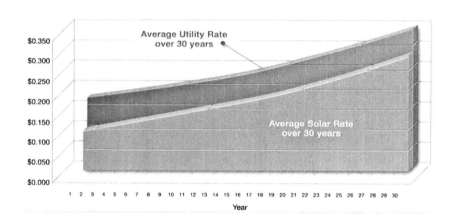

AN EXAMPLE OF PPA FINANCING:

The following example is from a 1 MW Southern California project. While the numbers below are not representative for all PPAs, they still can be used to show what is possible with a PPA.

Criteria for a Low Cost PPA
Stable Host

Not all solar systems or hosts are viewed equally by a PPA provider. The key evaluation criterion for a PPA provider is stability. Ideally, they want the host customer to have been in business for at least 5 years with strong credit. They also want the business to own their building, but

PPA Example – 1 MW

Estimated Savings Analysis: SCE GS2 TOUa Electricity Rate
Forecasted Savings and Cash Flow Summary - Current Solar Rebate

Year	Electricity Cost Without Solar	Equivalent Solar Power Cost	Savings With Solar	Annual % Savings
1	$396,000	$285,600	$110,400	28%
2	$413,664	$296,214	$117,450	28%
3	$432,116	$307,222	$124,894	29%
4	$451,391	$318,639	$132,752	29%
5	$471,526	$330,480	$141,046	30%
6	$492,559	$342,762	$149,797	30%
7	$514,530	$355,500	$159,031	31%
8	$537,482	$368,711	$168,771	31%
9	$561,457	$382,413	$179,044	32%
10	$586,501	$396,624	$189,877	32%
11	$612,663	$411,364	$201,299	33%
12	$639,991	$426,651	$213,340	33%
13	$668,539	$442,507	$226,032	34%
14	$698,360	$458,951	$239,408	34%
15	$729,511	$476,007	$253,504	35%
16	$762,052	$493,697	$268,355	35%
17	$796,044	$512,044	$284,000	36%
18	$831,553	$531,073	$300,480	36%
19	$868,645	$550,809	$317,836	37%
20	$907,392	$571,278	$336,114	37%
Total	**$8,206,289**	**$5,599,644**	**$2,606,645**	

some PPA providers will still enter an agreement with a long term lease in place.

Projects at Least $1,000,000 in Value

There are substantial costs associated with each PPA deal due to all the legal, tax structure, and engineering fees. With a larger project, those fixed costs are spread throughout more solar and have less of an impact on the PPA price.

Excellent Solar Location

The more electricity a given solar installation can produce, the lower the resulting PPA price will be. The reason is that a PPA provider will purchase the solar system and require a certain return on investment for the PPA term. The more a system produces over that term, the lower the resulting kWh price will need to be to cover their investment return requirements.

HOW TO RECEIVE A QUOTE FOR A PPA

There are a number of PPA firms available. Below you will find a short list of several reputable PPA providers, but there are many others available. Not all providers will offer the same price for a PPA for a given project so it is good to ask for a PPA quote from more than one. In order to receive a PPA quote, the PPA provider will need to know:

- Location of the solar project
- Product specifications (panels, inverters, and racking)
- Installation parameters (system size, azimuth (S, W, etc), Tilt)

Keep in mind that many solar system integrators can evaluate your facility and develop the project details and source a PPA provider for you. Many solar system integrators can handle everything in house and also some PPA providers will only work with a few solar installers for their projects because they are confident the installation will be quality.

PPA PROVIDERS

MMA Renewable Ventures: http://www.mmarenewableventures.com/
Recurrent energy: http://www.recurrentenergy.com/
Solar Power Partners: http://www.solarpowerpartners.com/
Tioga energy: http://www.tiogaenergy.com/

REPUTABLE SOLAR INTEGRATORS

REC Solar: http://www.recsolar.com/cm/Home.html
Sunpower: http://www.sunpowercorp.com/
Sun Edison: http://www.sunedison.com/

Chapter 12

Key Tips for Energy Solar Projects from Solar Power Association

Mike Taylor, Director of Research
Solar Electric Power Association (SEPA)

What would you say if I told you it's possible to have hundreds of solar panels installed on your building, offsetting a sizeable portion of your annual electricity use, with zero-down and an immediate payback? In addition, you won't take the risk of ownership and someone else will install and maintain it for 10-20 years. What's the catch? Ask Wal-Mart or Kohl's, as well as Macy's, Staples and Whole Foods, all of whom have taken advantage of a new business model at over 100 store locations.

Sound too good to be true? It's not... in the right situation.

Contrary to the common vision of solar panels on a residential home, the commercial sector represents the majority of the installed photovoltaic capacity in the solar industry, and has for some time, even as the overall industry has grown rapidly (Figure 12-1).

The solar energy services (SES) model is characterized by a third-party solar company owning and operating the solar system and selling solar electricity to the building owner, generally at a rate which is competitive compared to their existing prices. The building owner(s) are essentially buying solar electricity instead of a solar system, and offsetting some portion of their annual electricity use, while continuing to purchase the remainder from their traditional electric utility. Within the commercial sector, the SES model is rapidly becoming the business structure of choice for solar power, supplanting direct ownership of the PV system by the site operator, and is forecast to rapidly become the option of choice going forward (Figure 12-2).

The following discussions will provide some basic information about the solar services model with the specific target audience of potential building managers who may be interested in exploring it further.

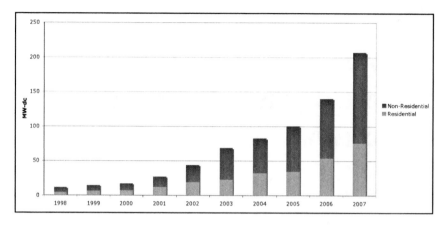

Figure 12-1. Annual Installed Photovoltaic Capacity by Sector (1998-2007)

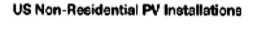

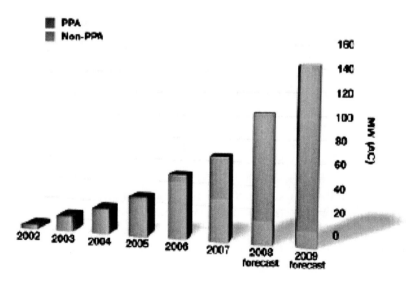

Figure 12-2. Non-residential Photovoltaic Market by Business Model. PPA = Power Purchase Agreement (Solar Energy Services); Non-PPA = direct ownership. *Source: Greentech Media, "Solar Power Services" 2008*

WHAT IS THE SOLAR ENERGY SERVICES BUSINESS MODEL?

A solar energy services contract can be an attractive option for large, commercial building owners/operators. It offers the option of paying for solar electricity instead of solar panels. A solar performance company will install, own, operate and maintain a solar electric system on your building for a period of up to 20 years. In return, you enter into a power purchase agreement (PPA) to purchase all of the solar electricity output from the system at a long-term, fixed price per kilowatt-hour that is competitive with your current electricity rates - in some cases up to 20% lower. The solar company acts as a solar utility and your electric bill is now divided between paying two companies, perhaps 70% to your traditional electric utility at their existing and future rates, and 30% to your new solar utility at a long-term fixed rate. At the end of the contract term, the arrangement can be renewed or terminated (and the equipment removed), with an optional purchase arrangement as well.

How Do They Do It?
The solar company maximizes the economies of scale in their overall portfolio and minimizes the individual project transaction costs. Unlike owning and installing your own individual project, solar performance contractors can, among other things, do the following:

• Raise the necessary capital and develop financing arrangements

• Fully utilize federal, state, and/or utility incentives

• Negotiate favorable pricing arrangements for solar products with aggregated volumes

• Understand state-specific requirements

• Manage utility electric rate structures and the interconnection processes

• Fully understand a rapidly changing technological, political, and regulatory business environment

Why Do They Do It?
For profit. These are conventional businesses—usually major financing and investing institutions—which exist to create financial value for their investors. These installations are profitable because of the available

federal, state, and/or utility incentives, but the business structure can capture economies of scale and portfolio optimization that individual non-solar businesses cannot.

What Do You Need to Know?

This business model isn't available everywhere for everyone, and is most active in southwestern and northeastern states. The following are three important considerations to take into account to determine if your business may be well-suited for SES, currently or in the near future:

1.   *Solar Resource*

Solar resource refers to the amount of available sunshine in any given location and therefore is determined by geographic location and associated weather patterns. There are some common misconceptions about solar resource, but it is one of the first things to consider. However, solar resource unnecessarily overshadows other considerations and is not the determining factor for whether to investigate SES further.

2.   *Electricity Prices*

A location with a great solar resource but lower electric rates may be less favorable than one with an average solar resource and high electric rates, i.e. New Mexico and Maryland could be equal when both resource and price are considered. Ultimately, a solar system is offsetting the retail price of electricity, not any particular electricity technology.

3.   *State Regulatory and Incentive Policies*

State policies are arguably the biggest driver of the SES model, even more than electricity rates and solar resources. Together, the triad of resource, price, and policy paints the solar business environment picture (Figure 12-3).

As an example, Germany has a solar resource comparable to Alaska (the lowest in the United States), has high electricity rates, and as a result of policy, leads the world in solar development, installing 10 times the amount of solar installed in the U.S. in 2006. After California, New Jersey is the second largest solar market in the U.S. for similar reasons.

New state solar policies often follow electricity price increases and/or volatility. Options such as energy efficiency, demand-side management, and solar energy often become part of resulting energy policy packages. Many of these policy initiatives are driven not just by the desire for the

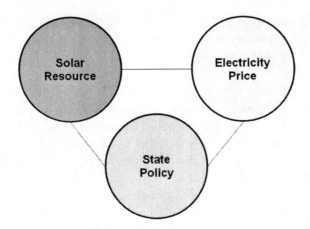

Figure 12-3. Solar Assessment Triad

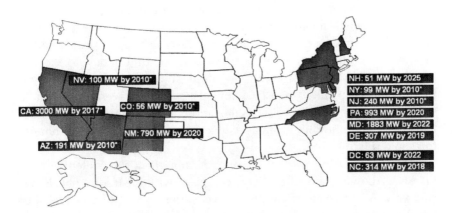

Figure 12-4. States with solar specific policies. *Source: Interstate Renewable Energy Council (derived and adapted; January 2008)*

specific solar installations, but also by the long-term value-added benefits of being an early adopter in a new technology area, i.e. research, jobs, manufacturing, etc.

At the policy level, the development of solar markets within any particular state is influenced by rules governing both that state's solar market and its respective system integration. Thirteen states have policies that specifically encourage more active solar markets (Figure 12-4). Many states have general renewable electricity requirements, i.e. 10% by 2015, but the thirteen states in figure 4 specifically carve out a solar requirement.

These long-term solar policies usually translate into the development of solar specific monetary incentives by the state or utility, which make it attractive for third-party solar developers to move into a particular state market.

California and New Jersey are the most active solar performance markets, but the other 11 states are in various stages of new development. A new solar state can easily jump into the solar policy market, as Maryland, Pennsylvania and North Carolina have recently done, but there is usually a 2-4 year ramp-up period between when the policy is enacted and when the market begins to become active. However, unanticipated state budget constraints or policy changes can also slow active markets, as has been seen in New Jersey as well. The one exception to the solar policy requirement is Hawaii, where high electricity rates and the solar resource make the market inherently attractive.

At the system integration level, state or utility net metering policy largely determines the maximum solar system size that can be installed at any one location. Net metering allows the customer's generation to offset onsite consumption at retail rates, and utilize the traditional electric grid as the balancing agent for minute-to-minute and day-to-day fluctuations in sunlight. A cap of less than 250 kilowatts in size (the average household system is around 3-5 kilowatts) significantly limits commercial solar markets, and within the United States only fourteen states meet this minimum threshold (Figure 12-5).

Size limits alone are not the only defining factor for potential solar markets. Allowing a third-party to "compete" with the traditional electric company by selling electricity to their customers is an additional legal issue that may need to be addressed. And the interconnection process can affect the level of time and information provided to the utility to ensure safe operation.

Beyond these three basic assessment criteria, there are additional considerations that make your particular location more or less attractive to a SES developer:

1. *System Size*

    If you meet the first three criteria (resource, electricity price, and policy), a large building with a flat roof, a large ground area (such as a parking lot or undeveloped area), or multiple buildings/lots with a similar aggregate is the next consideration. Although the size threshold will vary, a minimum of 50 kilowatts (5,000 – 10,000 ft^2, depending on the technol-

Figure 12-5.
Net metering
capacity limits for
individual solar
systems (kilowatts).
*Source: Interstate
Renewable Energy
Council (derived and
adapted; January
2008)*

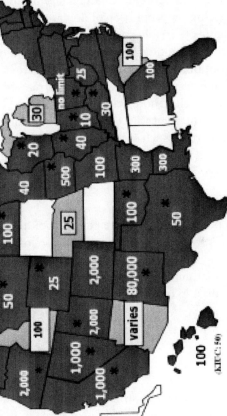

Commercial Net Metering Limits

October 2008

VT:	250
NH:	100
MA:	2,000*
RI:	3,500*
CT:	2,000*
NY:	2,000*
PA:	5,000*
NJ:	2,000*
DE:	2,000*
MD:	2,000
DC:	100
VA:	500*
NC:	100*
FL:	2000*

State-wide net metering for all utility types

* State-wide net metering for certain utility types only (e.g., investor-owned utilities)

Net metering offered voluntarily by one or more individual utilities

Source: Interstate Renewable Energy Council

ogy) is required, with larger systems and/or multiple buildings potentially receiving more competitive bids and pricing. Currently, solar markets are supply constrained and smaller systems and/or smaller solar market states may have difficulty attracting interest, but forecasts over the next 2-5 years indicate a flip in the market dynamics (see #7, below).

2.   *Electricity Rate Structures*
      Your electricity rate structure matters significantly as well. Electric rates that contain a higher proportion of demand charges (fixed costs) and lower energy charges (variable costs) are less attractive for solar energy since it is primarily offsetting energy use over time. It may be possible to analyze the rate options and choose one that is more heavily weighted to energy consumption. Several utilities in California are in processes with state regulators to develop more solar friendly rates. You might also consider changing to a different rate based on the resulting change in your daily consumption profile such as a time-of-use rate that has higher charges during the day and is lower at night. But a specific analysis of your situation should be done in both cases.

3.   *Financial Strength*
      The power purchase agreements for SES contracts are long term arrangements of 10-20 years. There are significant costs to setting up (and removing) a solar system, both technological and legal, and both parties want to ensure a long-term working relationship. The solar performance company will evaluate your financial stability, and given the growth of new solar companies without significant track-records, you should do the same to make sure they will be around to manage your system over the contract length.

4.   *Electricity Consumption*
      Your solar performance contract may offer an immediate price break from your current utility rates, for some percentage of your annual electricity use; 20-40% is common. But you are not necessarily reducing your electricity use, only changing where you purchase some portion of it. If electricity prices decline, your solar contract becomes less attractive, but over 20 years the risk is low. In addition, solar costs could decline over the next 5-10 years to the point where self-ownership is more financially attractive. Owning your own system may be preferable, but for many companies, the model will retain attractiveness based on the lower risk of

someone else maintaining the system over long periods.

5.   *Renewable Electricity*
     Your building (and probably your budget) doesn't care where the electricity comes from, renewable or otherwise. But as renewable energy or climate change issues become more prevalent, different methods of certifying, verifying, auditing, buying and selling renewable electricity are emerging. The most common accounting instrument is known as renewable energy credits (RECs), which are certified, tracked and traded through electronic trading systems operated by state, private, or non-profit organizations. As renewable electricity is produced, the electricity is immediately consumed, but virtual RECs for each megawatt-hour of electricity are created and registered in a government sanctioned tracking system, something akin to stock certificates. The tracking system verifies both generation and trades between buyers and sellers, but generally isn't managing financial transactions, which occur through private brokers and markets.

     Your contract with the solar performance provider will specify who owns the RECs, and they could be divided both proportionally and over time. If you don't own the RECs, technically you are buying non-renewable electricity and marketing efforts around your solar project would need to reflect this REC ownership status. For instance, if you didn't own your RECs you could claim to have solar panels on your facility, but not that your product or facility is 25% solar powered or offsets a certain number of tons of carbon dioxide each year. The RECs and the claim to their environmental benefits are owned by someone else and may have been sold elsewhere. The tracking system is in place to ensure that the RECs are not double counted, and can be retired toward renewable or carbon policies by utilities, generators or other stakeholders. If you want to use the project to reduce your company's carbon footprint (voluntarily or otherwise), owning the RECs would be required.

     REC ownership can be structured any number of ways, but keeping your RECs will influence the contract price. You should assess the costs and benefits, as well as your current and future goals in the negotiation process. Currently, solar REC markets are very active in New Jersey (and emerging in other eastern solar states), but without a solar requirement in a state, solar RECs would be priced similarly to wind RECs or energy efficiency carbon offsets, since they are least cost options in their respective arenas. Generally speaking the REC markets are just moving past the infancy stage and would require either a broker or internal resources to understand and navigate on your own.

6. *Solar Technology*

For the last 50 years, all commercially available solar technologies were based on highly purified silicon. In the last five years, this has changed rapidly as thin film technologies, i.e. tens of microns thick instead of hundreds of microns thick, are gaining significant market share from four new available technology types. Thin films offer lower efficiencies and require greater roof space, but generally at lower cost. While not a major part of your decision process as an SES customer, recognizing basic differences is important. For instance, a crystalline silicon solar panel project may be slightly more expensive than a competing thin film project on a contract price, but given a fixed roof space, a large sized system could be installed with the traditional silicon system. Given its higher efficiency, the silicon-based panels might offset 40% of your annual electricity use rather than 20% for thin films, i.e. silicon has higher energy production per square foot.

7. *Small Systems/Small Markets*

Silicon still makes up 90% of the solar supplies, and with a shortage that began in 2005 and will continue through 2009, companies can afford to be less aggressive in reaching out to smaller individual system purchasers and/or states with smaller markets. A 50 kilowatt system certainly has less market appeal than a 500 kilowatt system. Similarly, a 250 kilowatt system in California, where the market infrastructure is more mature, may have more appeal to a developer than a 500 kilowatt system in Kansas, where little solar activity is taking place. Multiple locations with a larger aggregated size will hold similarly commensurate appeal. However, over the next few years (around 2010), major global solar manufacturing capacity expansions will relieve this situation and companies will be more aggressive about finding new customers and markets.

8. *Flexible Terms*

The terms of the contract, whether length, cost (fixed, inflation-adjusted, escalating, declining, etc), RECs, or buy-out options are all generally negotiable as long as the net-present value of the project is similar to the SES company. Understanding your goals for the project—whether strictly electricity bill reductions, environmental impact reduction, or some combination will affect how you approach the options. But those goals should be clearly delineated internally as you move forward with negotiations.

HOW DO I BEGIN THE EVALUATION PROCESS?

If your company issued an RFP for a solar performance contract spread over 20 buildings totaling 10 megawatts in California, a strong number of SES companies would respond with proposals. If your company issued an RFP for one 50-kilowatt project in Kansas, it is unclear what the response would be. How to evaluate your situation in between is not necessarily clear. You could contact one of the SES companies directly, but it would be smart to understand a little about your particular solar market framework and system potential first.

- State Market – First, assess your state's current and future market. Are solar specific and basic net metering policies in place? Will they be in the near future? Are solar incentives available? Visit www.dsireusa.org for more information.

- System Size – Second, assess your potential installation size. How much roof or ground space is available at how many locations? Ballpark 100 ft2/kilowatt for silicon panels and 175 ft2/kilowatt for thin film panels.

In the near-term, companies should further investigate the SES model if they are located in California, Hawaii, or New Jersey and have adequate roof space, particularly those with multiple locations that can aggregate them into a larger contract. While the market is currently limited, new state markets are emerging in Arizona, Colorado, Delaware, District of Columbia, Maryland, Nevada, New Mexico, New Hampshire, New York, North Carolina, and Pennsylvania. As new global manufacturing supplies increase, the supply/demand dynamic will turn in the buyers' favor, making companies more aggressive in working on smaller projects in these new market states.

You can also utilize FindSolar.com (www.findsolar.com) to locate commercial "solar pros" who offer a power purchase agreement (PPA) in your location. The number of companies listed, especially those with a greater portfolio of projects, is a good indicator of your state's solar market activity. Cross referencing the list against companies sponsoring or exhibiting at the annual Solar Power International conference (www. solarpowerinternational.com) would give you a good idea of the larger or more active national and regional companies. The conference would also be well worth attending as your project becomes more concrete, especially for larger investment plans.

Chapter 13

How Green Projects Affect Stock Prices

How High? As much as 21.33% within 150 days of an Announcement!*

John R. Wingender, Jr. and Eric A. Woodroof†

EDITOR'S COMMENTS:
This chapter was published in 1997, yet it still receives great attention and is still being translated into new articles around the world… the conclusions are memorable. Based on today's eco-savvy consumers, a follow-up study would be very interesting to conduct. That is an invitation…

ABSTRACT

When selecting projects under a limited budget, a firm should implement the project that will return the most value. Ultimately, firm value is measured by stock price, which can be impacted when the firm makes a major announcement. This chapter shows that announcements of energy management projects correlate with a 21.33% increase in stock price within 150 days of the announcement. This increase is in addition to the risk-adjusted return the firm would normally experience. For example, during a "bull market" a firm's expected return was 10%. After the announcement, the return would increase by 21.33%, for a net return of 31.33%. These results suggest that investors react positively to energy management projects. This outcome demonstrates one more strategic incentive for firms to implement energy management projects.

---

*Published in Strategic Planning for Energy and the Environment, Vol. 17(1), pp. 38-51, 1997
†To whom all correspondence should be addressed.

INTRODUCTION

The potential for increased profits via cost-reducing Energy Management Projects (EMPs) exists in nearly all firms. However, when allocating capital, priority is often given to revenue-enhancing projects, such as starting new product lines or joint ventures. Frequently, these projects are perceived to be superior to EMPs, even though they may yield the same increased profit and present value. A justification is that revenue-enhancing projects are more likely to attract publicity and investor attention. Investor speculation and reaction to announcements can increase the firm's stock price. Most EMPs do not generate as much publicity as joint ventures or new product lines.

If "publicity-gaining" potential is a decision factor during project selection, then a new product line or joint venture would usually be selected over an EMP. But is this a fair comparison? There has not been any research to determine if an EMP announcement increases a firm's stock price. In theory, it should because most EMPs increase profits (via cost reduction instead of increased revenues). From a cash flow perspective, an EMP is equivalent to any other profit-enhancing project.

This chapter seeks to determine whether an EMP announcement correlates with an abnormal increase in a firm's stock price. If such announcements positively impact stock price, then the firm has one more incentive to implement EMPs.

LITERATURE REVIEW

The purpose of this literature review is three-fold:
1.   To demonstrate that EMPs are credible investments, with relatively low risk;
2.   To present some background on stock price reaction to announcements of typical capital investments; and
3.   To show that abnormal increases in stock prices from EMP announcements have not been measured.

Public announcements (such as mergers, joint ventures or new product lines) correlate with abnormal stock price returns.[1,2] When a firm announces a joint venture (or other revenue-enhancing project) it is trying to attract publicity, which can raise the stock price based on expected future

profits. However, since such projects can also be unprofitable, the anticipated cash flows are at risk.

When firms implement EMPs, they also expect improved profits
by becoming more cost-competitive. EMPs and equipment replacement
projects usually have more predictable cash flows (less risk) than many
other types of capital investments, especially new product lines or joint
ventures.[3] Today, the risk from most EMPs is so low there are many third
party lenders who are eager to locate and finance EMPs.[4] In 1995, leasing (which does include third party leasing and performance contacting)
accounted for nearly one third of all equipment utilization.[5] Thus, EMPs
and other facilities improvement projects are recognized as credible investments, however they are frequently put on the "back burner" relative
to revenue-enhancing projects.

Maximizing stock price should be a goal of the corporation. Increasing productivity, offering new product lines, and increasing profits are
examples of tangible factors that can increase the firm's stock price. However, stock price may also increase due to intangible factors; such as investor speculation and reaction immediately following an announcement.
Executives may incorporate this investor reaction when deciding which
projects to implement.

Although investor reaction has not been assessed for EMP announcements, there has been some research in this area. It has been shown that
firms increasing expenditures on general facility and equipment improvements had a 1.98% abnormal stock increase immediately after the announcement.[6] Announcements of joint ventures correlated with a positive
abnormal return of 1.95% over a twenty-one day interval, (-10 days to +10
days) centered on the announcement day (day 0).[7] However, not all joint
ventures correlate with positive abnormal stock returns.[8]

The correlation between EMP announcements and stock price has
not previously been investigated. This chapter will examine whether EMP
announcements correlate with positive abnormal stock price returns. If
they do, then perhaps the capital budgeting process should incorporate
this benefit.

Methods
    Using the Nexis/Lexis database, the World Wide Web, and other re-
sources, a search for EMP announcements resulted in over 5,500 citations.
Of the 5,500 citations, only 23 announcements fit the following criteria:
1.   The firm announcing the EMP was publicly traded and its returns

were available on the data files of the Center for Research in Security Prices (CRSP).

2.   The announcement was the first public information released about the EMP.

3.   The EMP must be large enough to represent a significant investment for the company. For example, if a large fast-food chain was announcing an EMP at only one restaurant, it was deleted from the sample. However, if the EMP was company-wide, it would be included in the sample.

4.   The announcement was made in a major US newspaper, news wire service, or a monthly trade magazine between 1986 and 1995.

The 23 announcements which represented the "Complete Sample," consisted of several sub-samples:

•   16 Announcements made within daily newspapers or via electronic wire ("Daily")

•   3 Announcements within monthly trade magazines ("Monthly")

•   4 Announcements of post-implementation results ("Post-Implementation")

An additional sub-sample was created ("Daily + Monthly") to maximize the sample size of pre-implementation announcements.

For each announcement, the firm's name, CRSP identification number and announcement date were entered into the Eventus computer program, which tracks daily stock price performance for every U.S. stock listed within the CRSP.[9] Eventus uses a common "event-study" methodology to calculate abnormal stock returns, and indicates statistical significance levels. Statistically, a null hypothesis (H_o) was proposed and tested against an alternative hypothesis (H_a):

H_o = EMP announcements have no impact on stock price.

H_a = EMP announcements have a positive impact on stock price.

Analysis Intervals

Usually, a firm's stock price performance is analyzed over several time intervals around the announcement date. Often, a stock price improvement can be noticed between the day of the announcement (day 0) and the next trading day (day 1). In this interval, the range is represented by the following notation: (day 0, day 1) or (0,1). Another typical interval for event-study analysis is a two-day interval, one day before the announcement to the announcement date (-1,0).[10]

EMP announcements frequently appear in monthly trade magazines, which are distributed to readers on different days in different geographic locations. In this case, an exact announcement date can not be determined. Thus, analysis of the stock performance over a wider interval is appropriate. The (-10, 10) interval represents the period at which EMP announcements would most likely be noticed through monthly trade magazines. In addition, since EMP announcements may not capture as much publicity as other announcements, they require a longer period for the market to "learn" about them.

The (-10,10) interval is useful for identifying if an abnormal stock price increase correlates with an EMP announcement. However, to observe the long-term stock impact, the sub-samples were analyzed over additional intervals, such as (1,100), (1,150), etc.

Applying the aforementioned hypothesis tests to the sub-samples yielded the results that are presented in the following section of this chapter. A more detailed explanation of the "event-study" methodology (statistical analysis) is included in the Appendix.

Results

Tables 13-1 and 13-2 present the short-term and long-term abnormal returns. The returns are categorized by interval around the announcement date. The level of significance at which H_0 was rejected is also indicated.

The Daily + Monthly sub-sample is the most appropriate sample because it is the largest sample possible that excludes Post Implementation announcements. The Post Implementation sub-sample is substantially different in nature because it represents firms announcing cost savings (increased profits) from projects already implemented.

Using the Daily + Monthly sub-sample, EMPs correlate with a 3.90% increase in stock price, measured from ten days prior to the announcement to ten days after the announcement, (-10, 10). The level of significance was 0.01. See Table 13-1.

Table 13-2 shows the long-term performance, where the Daily + Monthly sub-sample correlated with a 21.33% abnormal return over the (1,150) interval, at the 0.001 significance level. Figure 13-1 is a graphical illustration of the abnormal returns over the long-term interval.

Table 13-1. Abnormal Return of Firms Announcing Energy Management Projects

Samples	# of Firms in Sample	Day Range From Announcement Date			
		(-5,5)	(-10,10)	(-15,15)	(-20,20)
Complete Sample	23	2.21%, *	2.80%, **	1.06%, *	1.92%, *
Announcements from Daily Newspapers	16	2.89%, **	3.75%, ***	2.31%, **	3.18%, **
Announcements from Monthly Magazines	3	2.31%	6.91%	8.23%	8.50%
Post Implementation Announcements	4	-0.57%	-2.70%	-5.86%	-7.12%
Daily + Monthly Subsamples Combined	19	2.43%, *	3.90%, ***	2.83%, *	3.64%, **

Significant at: * =0.10
              ** = 0.05
              *** = 0.01

Table 13-2. Long-Term Abnormal Return of Firms Announcing Energy Management Projects

Samples	# of Firms in Sample	Day Range From Announcement Date					
		(1,100)	(1,120)	(1,150)	(1,200)	(1,220)	(1,240)
Complete Sample	23	8.23%, *	11.00%, **	13.93%, ***	12.09%, **	13.75%, **	12.71%, **
Announcements from Daily Newspapers	16	10.67%, **	16.29%, ***	17.66%, ***	13.38%, **	17.10%, **	17.11%, **
Announcements from Monthly Magazines	3	27.05%	21.38%	31.54%	44.89%	42.44%	53.02%
Daily + Monthly Subsamples Combined	19	14.07%, **	18.03%, **	21.33%, ****	21.73%, ***	23.42%, ***	24.95%, ***

Significant at: * = 0.10
              ** = 0.05
              *** = 0.01
              **** = 0.001

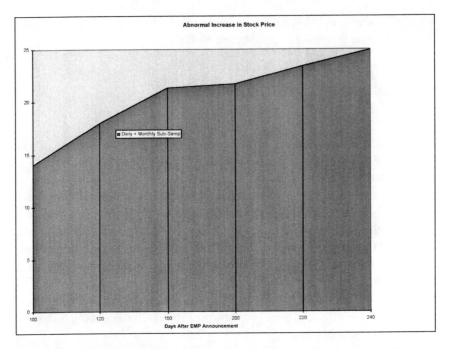

Figure 13-1. Long-Term Abnormal Return of Firms Announcing Energy Management Projects

CONCLUSION

The results from this study indicate that EMP announcements correlate with significant abnormal increases in a firm's stock price. On average, an EMP announcement correlated with a 21.33% abnormal increase in the firm's stock price. This increase was experienced from the day after the announcement to 150 days after the announcement. This increase is in addition to the risk-adjusted return the firms would normally experience. For example, during a "bull market" a firm's expected return was 10%. After the announcement, the return would increase by 21.33%, for a net return of 31.33%. Because these EMPs were announced by a diverse group of firms at various periods over a ten-year time span, the significance of these results is impressive. In other words, the EMP is probably the only event that all firms within the sample have in common.

From these results, it appears that shareholders recognize EMPs as low-risk investments that should increase profits and add value to the firm. With the new information presented here, firms may have an additional strategic incentive to implement EMPs.

DISCUSSION AND RECOMMENDATIONS
FOR FURTHER RESEARCH

Despite the small scale of this study, the significance of the results is impressive. This study could serve as a "first-step" to understanding investor reaction to EMP announcements. Additional studies with increased sample size and greater stratification would yield more information.

It is interesting to note that detailed cost savings estimates were not always included in the EMP announcements. *For example, many firms simply announced that they were going to retrofit a portion of their facilities, without an estimate of dollar savings.* Perhaps more detailed information was released after the announcement date, triggering greater stock price increases in the long-term intervals. However, it is more likely that shareholders associate EMPs as effective profit enhancing projects that are almost always good for the bottom line.

It would be interesting to determine if there is a relationship between an EMP's potential profits and the value of the abnormal return. The value of the abnormal return should be related to the amount of increased profit from the EMP. Identifying these values could indicate whether the investor reaction is proportional to the potential added value of the EMP. Calculating these values would require additional information about each firm as well as each project. This could be a focus of additional research.

It was recognized that the type of EMP could influence the magnitude of the abnormal stock increase. Thus, the Complete Sample was further stratified into two sub-samples: EMPs that were Lighting Retrofits and EMPs that were installations of other types of Energy Efficient Equipment (such as HVAC upgrades, chiller upgrades, etc.). The Eventus software analyzed both sub-samples, and the Energy Efficient Equipment sub-sample correlated with a 1.42% increase over the (-10,10) interval, at the 0.01 significance level. The Lighting Retrofit sub-sample yielded no significant returns over the (-10,10) interval. However, because this sub-sample only contained seven firms, this comparison needs to be re-evaluated with a greater sample size. In addition, this analysis was tainted because the post-implementation announcements were included.

It was also recognized that the finance method for each EMP could influence the magnitude of the abnormal stock increase. Since off-balance sheet financing (leasing) is common for EMPs, a comparison was made between EMPs that utilized leasing versus EMPs where the equipment was purchased by the firm (and the debt was carried on their balance sheet). Assuming that stock analysts frequently look at balance sheets to assess a company's performance, it is reasonable to hypothesize that off-balance sheet financed EMPs would correlate with higher abnormal stock returns than EMPs where equipment was purchased by the firm. The Complete Sample was further stratified and analyzed by the <u>Eventus</u> software. The EMPs that were purchased directly by the firm did show a 3.74% increase over the (-10,10) interval, at the 0.01 significance level. The sub-sample of leased EMPs yielded no significant returns over the (-10,10) interval, although this sub-sample included only three firms. Again, the samples were tainted with the post-implementation announcements. Therefore, this comparison needs to be re-evaluated with a larger sample size, and with the post-implementation announcements removed.

Although the Post-Implementation sub-sample was small, it did not yield any significant positive abnormal returns. In fact, the returns were negative, although not significant. This is intriguing because a post-implementation announcement is basically a statement of increased profits already realized. A more extensive study on the effects of Post-Implementation announcements could reveal if these types of announcements yield different abnormal returns than announcements prior to implementation.

All of the sub-samples mentioned in this Discussion Section should be analyzed over a longer time interval. In addition, increasing the sample size would also improve the validity of the results.

References
McConnell, J. and Nantell, T. (1985), "Corporate Combinations and Common Stock Returns: The Case of Joint Ventures," *The Journal of Finance* 40(2), p. 519-536.
Pettway, R. and Yamada, T. (1986), "Mergers in Japan and their Impacts on Stockholders' Wealth," *Financial Management* 15(4), Winter 1986, p. 43-52.
Pohlman, R., Santiago, E., and Markel, E. (1988), "Cash Flow Estimation Practices of Large Firms," *Financial Management* 17(2), p. 71-79.
Zobler, N. (1995), "Lenders Stand Ready to Fund Energy Projects," *Energy User News* 20(3), p. 19.
Sharpe, S. and Nguyen, H. (1995) "Capital Market Imperfections and Incentive to Lease," *Journal of Financial Economics*, 39(2), p. 271-294.
McConnell, J. and Muscarella, C. (1985), "Corporate Capital Expenditure Decisions and the Market Value of the Firm," *Journal of Financial Economics*, 40(3), p. 399-422.
NOTE: Abnormal return was measured over a two-day announcement period that encompasses the day on which the announcement appeared in print, plus the previous day (-1,0). The stock performance over a twenty-one day interval (-10,10) was not

reported. This abnormal return value can be different than the value over the two-day interval. The result was different from zero at the 0.01 statistical significance.

McConnell, J. and Nantell, T. (1985), "Corporate Combinations and Common Stock Returns: The Case of Joint Ventures," *The Journal of Finance* 40(2), p. 519-536.

Lee, I. And Wyatt, S. (1990) "The Effects of International Joint Ventures on Shareholder Wealth," *The Financial Review* 25(4), p. 641-649.

Cowan, A. (1996), *Eventus Version 6.2 Users Guide*, Cowan Research.

Brown, S. and Warner, J. (1985) "Using Daily Stock Returns," *Journal of Financial Economics*, 14(1), p. 3-31.

APPENDIX

EVENT-STUDY METHODOLOGY

An "event study" is a popular analysis tool for analyzing stock price reactions to particular events. In this study, the "event" is the announcement of an EMP by one of the sample firms. The event date is the first trading day that the market could react to the announcement. Calculating risk-adjusted abnormal returns on and around the announcement date tests the impact of EMP announcements on stock price.

For this study, we use the market model event-study method and test the results for significance with the standardized residual method. The market model event-study method uses a linear regression to predict stock returns; then it compares the predicted value to its actual returns.

The abnormal return (ABR_{jt}) is the difference between the actual return (R_{jt}) on a specific date and the expected return ($E(R_{jt})$) calculated for the firm on that specific date. The expected return is calculated using the parameters of a single index regression model during a pre-event estimation period. The regression model parameters are determined by the following equation:

$$R_{jt} = a_j + b_j R_{mt} + e_{jt}$$

where

R_{jt} = the return on security j for period t,

a_j = the intercept term,

b_j = the covariance of the returns on the jth security with those of the market portfolio's returns,

R_{mt} = the return on the CRSP equally weighted market portfolio for period t, and

e_{jt} = the residual error term on security j for period t.

To calculate the market model parameters (a_j and b_j), a 220-day estimation period was used that begins 260 days before the announcement date. For each sample firm, the event period begins 30 days before the announcement date and ends 30 days after the announcement date. The expected return ($E(R_{jt})$) is then calculated using the return on the market (R_{mt}) for the specific event period date:

$$E(R_{jt}) = a_j + b_j R_{mt}$$

The abnormal return (ABR_{jt}) for an event date is then calculated by subtracting the expected return (which uses the parameters of the firm from the estimation period and the actual market return for a particular date in the event period) from the actual return (R_{jt}) on that date. The equation is as follows:

$$ABR_{jt} = R_{jt} - E(R_{jt})$$

The average abnormal return (AAR_t) for a specific event date is the mean of all the individual firms' abnormal returns for that date:

where N is the number of firms used in the calculation.

The cumulative average abnormal return (CAAR) for each interval is calculated as follows:

The standardized residual method is used to determine whether the abnormal returns are significantly different from zero. The standardized abnormal return (SAR_{jt}) is calculated as follows:

$$SAR_{jt} = ABR_{jt} / s_{jt}$$

where

s_{jt} = the standard deviation of security j's estimation period variance of its ABR_{jt}'s.

The estimation period variance s^2_{jt} is calculated as follows:

$$\left[\sum_{k=1}^{D_j} \left(ABR^2_{jk} \right) \right] / (D_j - 2),$$

where

s^2_j =

R_m = the mean market return over the estimation period, and

D_j = the number of trading day returns (220) used to estimate the parameters of firm j.

Finally, the test statistic for the null hypothesis (H_0) that the $CARR_{T1'T2}$ equals zero is defined as follows:

where

$$Z^j_{T_1, T_2} = \left(1 / \sqrt{Q^j_{T_1, T_2}} \right) \sum_{t=T_1}^{T_2} \left(SAR_{jt} \right)$$

and

$$Q^j_{T_1, T_2} = (T_2 - T_1 + 1) \frac{D_j - 2}{D_j - 4}$$

Footnotes

1. McConnell, J. and Nantell, T. (1985), "Corporate Combinations and Common Stock Returns: The Case of Joint Ventures," *The Journal of Finance 40(2)*, p. 519.
2. Pettway, R. and Takeshi, Y. "Mergers in Japan and their Impacts on Stockholders' Wealth," *Financial Management, Winter 1986*, p. 43.
3. Pohlman, R., Santiago, E., and Markel, L. (1988), "Cash Flow Estimation Practice of Large Firms," *Financial Management*, Summer, p. 71.
4. Zobler, N. (1995), "Lenders Stand Ready to Fund Energy Projects," *Energy User News 20(3)*, p. 19.
5. Sharpe, S. and Nguyen, H. (1995) "Capital Market Imperfections and the Incentive to Lease," *Journal of Financial Economics*, 39 1995 p. 294.
6. McConnell, J. and Muscarella, C. (1985), "Corporate Capital Expenditure Decisions and the Market Value of the Firm," *Journal of Financial Economics, 40(14)*, p. 399. Abnormal return was measured over a two-day announcement period that encompasses the day on which the announcement appeared in print, plus the previous day (-1,0). The stock performance over a twenty-one day interval (-10,10) was not reported. This abnormal return value can be different than the value over the two-day interval. The result was different from zero at the 0.01 statistical significance.
7. McConnell, J. and Nantell, T. (1985), "Corporate Combinations and Common Stock Returns: The Case of Joint Ventures," *The Journal of Finance 40(2)*, p. 519.
8. Lee, I. And Wyatt, S. (1990) "The Effects of International Joint Ventures on Shareholder Wealth," *The Financial Review*, November, pp. 641-649.
9. Cowan, A. (1989), *Eventus Version 6.2 Users Guide*, Cowan Research.
10. Brown, S. and Warner, J. (1985) "Using Daily Stock Returns," *Journal of Financial Economics, 14(1)*, p. 3-31.

Chapter 14

Overcoming the Barriers that Delay "Good" Projects

EDITOR'S COMMENTS:
This is the final chapter and it contains application tips that would work with any project...

INTRODUCTION

It would be logical that cost-saving projects that also reduce an organization's environmental impact would be "no brainers" for management to approve. However, many "discretionary" projects have been delayed, taking a "back seat" to sales, production or other "must do" projects. It is hard to argue against that point of view, however there may be "more to the story," as outlined by the article below...the bottom line is that when all the benefits of a good "green" project are presented, it can be very hard to delay.

We define a good "green" project as one that has a 3-year payback (approximately 30% rate of return) and helps reduce carbon emissions or environmental impact. Reducing environmental impacts is common with energy-efficiency, waste-minimization and other "environmental projects." The projects can be simple or complex—requiring an engineer or manager to develop and supervise the project/program. However, engineers are *wasting their time* if the projects they so carefully develop are *not implemented* and deliver no value. This article refers to "good" projects as those with a 3 year payback or less. Why don't "good" projects get implemented? There are a variety of reasons and a few common barriers:

1. *Marketing* (under-marketing a project's value, or delivering to the message to the wrong person)

2.   *Education & Collaboration* (not expanding the value of a project)

3.   *Money* (not having a positive cash flow solution)

If a project can't beat these criteria, it probably won't be implemented anyway… so focus on the ones that will!

PROBLEM #1: MARKETING

People often ask me why marketing is first on the list. Answer: because NOTHING HAPPENS WITHOUT A SALE. For example, your first job (or your first date) began with you "selling yourself" on a resume or during an interview. In fact, the development of every product/service begins with someone selling a solution to some type of problem. Now, I am not saying that selling/convincing is "bad" or un-ethical. Convincing someone when it improves their lives is good and can be done with passion. When something (like an energy management project) is great, we should sell the benefits <u>with all the passion in the world</u>. *You would do the same when talking to your kids about "getting a good education," or "learning good manners."* Passion can also emerge from fear, such as from the chaos and the violence that occurs during an electrical blackout. <u>*Most of the time, humans are more passionate and action-oriented when they are at risk of losing something versus gaining something.*</u>

So, we must communicate in a way so the audience (the buyer or project approver) can understand the problem/pain that they are in now. After they agree that they are "in pain," then they will want to hear about potential solutions.

Attention

It starts with getting the buyer's attention on the problem, the pain it is causing and a sense of urgency to solve the problem. Only then will a solution seem to be logical. In addition, after they understand the problem/pain, they will be able to become passionate about the solution.

If you fail to get the attention of the approver, you are actually doing them a dis-service; they won't know they are in trouble and are wasting money. It is like allowing someone to bleed to death when they don't even know they are cut. So, don't be shy—you have a duty to perform.

Warning! Some approvers personalities' won't like to hear about problems/pain. Some approvers may "put their head in the sand" (like an ostrich) when problems are discussed. Don't blame them—it is their personality (which has strengths in other areas). Discover ways to communicate in a way that they will respond. FYI—it can take 7 impressions (explanations/presentations) before some people will agree on the problem and take action on a solution. Don't give up and don't be surprised or depressed when they don't take action after the first impression.

Below are a few examples of effective headlines[1] that can help get the attention of an approver. Feel free to use these in executive summaries:

- "How will the shareholders feel about us throwing money away every month?"

- "A way to make money while reducing emissions..."

- "What will we do with the yearly savings?"

- "We are paying for energy-efficiency projects, whether or not we do them!"

- "Guaranteed, high-yield investments..."

- "If you enjoy throwing money away every month, don't read this..."

- "4.6 Billion years of reliability... solar energy"

- This project could improve our stock price by over 20%!

- "Good planets are hard to find"

There are many other great proven examples that are available.[2] However, you can experiment by looking around for "marketing copy" in magazine advertisements, commercials, etc. *There is a reason they call it "copy"—some of the principles are thousands of years old, and they still work!* Just change the words to relate to your problem/solution. Try a few versions and test, test, test to see which ones are most effective. Go for it!

Benefits[3,4]

After you have their attention, be sure that you include compelling benefits that "take away the pain" that the audience is feeling. As engineers, we are good at mentioning the typical benefits:

- Saves energy, money, waste and emissions;
- Offsets the cost of a planned capital project;
- Improves cost-competitiveness, productivity, etc.;
- Is a relatively low-risk, high-profit investment that directly impacts the bottom line.

In today's green-minded economy, we could also demonstrate that "green" projects are a very effective marketing tool (which could get the client's marketing department behind your project), because these benefits have also been proven:

- Improves the client's "green" image;
- Differentiates the client from the competition;
- Introduces them to new markets, suppliers and clients;
- Helps them grow sales/revenue.

However, we should also mention the passionate, global and moral reasons behind a good "green" project:

- Slows global warming, reduces acid rain;
- Reduces mercury pollution, which allows us to eat healthy;
- Improves our national energy independence;
- Reduces security/disaster risks, etc..

Dollar values for these benefits can often be calculated and should be included in your proposal. To calculate the "green benefit equivalencies," such as "number of trees planted" (from reduced power plant emissions), see the "Money" section of this article.

The list above can be expanded, refined and optimized for any project. To build a list like the one above, one technique is: "WSGAT": "What is So Good About That?" Ask that question for every project feature, and you will develop a long list of passionate benefits. By the way—this approach has been used in TV sales and has helped sell billions of dollars of material[5]. *If they can sell this much junk on TV, we should have no problem selling green projects that are factually saving the planet!* Add the emotional benefits of going "green," and you will have a project that touches the hearts of leaders in your organization.

Call to Action

The "call to action" becomes easy and logical when all of the benefits have been quantified and they are aligned with the client's strategic objectives. Tell the approver what you want them to do and why. Be sure to include the "cost of delay" in your executive summary. Remind them that they are "in pain" and the project/solution will solve it. Visual aids can be helpful. *For example, during one presentation, buckets of dollars were shoveled out a window to demonstrate the losses that were occurring every minute. The executives were literally in pain watching those dollars fly away. They couldn't stand it, and they took action. It is OK to get creative and have some fun in your presentation!*

BUT WAIT... THERE IS MORE! "Configuring" your presentation can make the difference between immediate approval and further delay. There are many ways to "configure" or "package" your product/project so that it is IRRESISTABLE. One way to do this is to find a way to make a project's performance guaranteed or "risk-free." Another way is to separate (or add) one part of the project and introduce it as a "free bonus." Everyone likes a "FREE" bonus—it helps them understand that they are getting a "good" deal. *For example, on a recent "green," facility-related project, carbon offsets for a company's fleet were included as a free bonus. The bonus delighted the client and differentiated the project, (and it was extra value) yet the additional costs were less than $1,000.*

Engineers can be two, three or ten times more productive by developing sales and marketing skills. However, there is another reason for developing these sales/marketing skills: Your career! The skills you learn will be valuable to your organization (as well as other organizations). These skills are transferable to other industries too. So keep this in mind when you are investing in yourself... there will almost always be a fantastic pay-off.

Finally, there are two prerequisites that a buyer must see in you before any sale is made: "Trust" and "Value." As far as trust goes, it must be earned and once it is earned... it must be cherished. To accelerate the buyer's trust in you: be an advocate for the client and put their needs ahead of your own. Assume the role of their "most trusted advisor," (and then deliver). Value comes from applying knowledge, tools, resources, partners, etc., in the best way for the client, which is why education and collaboration is such an important component of success. This is discussed further in the next section. Be sure to read the sub-sections on reciprocal business agreements, Joint ventures and incentives/rebates... great ideas!

PROBLEM #2: EDUCATION & COLLABORATION

Knowing how to deliver the value is an area that requires continuous updating. Today, with the proliferation of energy/green technologies, it is impossible for one person to know all the ways to add value to a project. Green specialties are expanding every day. For example: energy efficiency, water efficiency, green janitorial, LEED[6], recycling, transportation, etc..

Learn all you can, then collaborate with other professionals (who are also actively learning) and the value available to your clients' increases exponentially. It is important to be open to new ideas and fresh perspectives in this process. *"Mind-Sharing" or brain-storming techniques can facilitate the process and maximize the number of useful ideas.*[7]

Fortunately, education is a low-cost investment. Collaboration and even joint-ventures/partnering can be done inexpensively as well and the returns can be huge!

Free Sources of Green/Energy Efficiency Education:
https://www.aeecenter.org/seminars/
http://www.eere.energy.gov/
http://www.ashrae.org/education/
www.usgbc.org
www.ase.org
www.energystar.org
http://greeninginterior.doi.gov

In addition, there are many innovative ways to bring more value to a project. Some include:
• Reciprocal Business Agreements
• Joint Ventures
• Free Tax and Utility Incentives/Rebates

Reciprocal Business Agreements:
For example: After presenting a $1,000,000 service contract for a global car rental company, the deal was sweetened with an agreement on our part to choose that car company while traveling, which generated over $1,000,000 in extra car rentals for them. To the client, they were getting an extra $1,000,000 in revenue by working with us versus the competitors. *What suppliers, partners, colleagues, professionals, etc. could you develop reciprocal business agreements? How could you help two clients (or a supplier) benefit from each other? How could you help them become more "green"?*

Another example: We helped client #1 supply green solutions to client #2. Both clients were extremely happy to generate more sales/save money. When it was time to approve our next round of projects, there was little resistance, because we had helped them earn/save far more than the costs of the proposed projects. This illustrates the value of being the "trusted advisor."

Joint Ventures

For example: A "green" travel agency gives 50% of its commissions back to its clients in exchange for their travel business.[8] The client can use this extra, free money to fund "green" initiatives or scholarships, or other social programs. The travel agency guarantees the lowest prices and easily doubles its business because it delivers more value to its clients via joint ventures.

Free Tax and Utility Incentives/Rebates

For example: In California, 50% of a solar project can be funded by federal and state rebates. Utility incentives lowered the installation costs even further. There are numerous free tax and utility incentives available, and some are discussed in the next section.

In addition to the options above, many utilities and third parties are offering "green power purchase agreements," which are essentially "wind and solar performance contracts." For example, if you want to put solar panels on your roof. A third party (often a utility or solar contractor) finances the project installation and then sells you the renewable energy produced from your roof (at a known price) for 15-25 years. So you get "green" power at no upfront cost, and a known future energy cost (lowers your risk to energy price volatility). The financier wins because the project will pay back their investment within 10 years and the rest is profit.

There are an unlimited number of creative "win-win" contracts available. However, before finalizing or even developing your solution, be sure that you understand the client's strategic and financial goals and align the value to support the client's larger objectives.

PROBLEM #3: MONEY

If you do a good job tapping into the passion behind the project and are satisfying the emotional, financial and other approval criteria, you should have enough benefits to get the project approved, especially if the project is above the client's MARR.[9] However, if your organization is capitally constrained, you can finance a project and have positive cash flow. *CFOs like positive cash flow projects!* On the contrary, cash flow constraints (not having the upfront capital to install a project) represent over 35% of the reasons why projects are not implemented[10].

Financing does not have to be complicated. In fact, financing energy efficiency/green projects can be very similar to your mortgage or car payment; Fixed payments for a length of time. However, with a "good" project, you can finance the project such that the annual savings are greater than the finance payments—which means the project becomes "cash flow positive" and does not impact the capital budget! This can allow the approver to move forward without sacrificing any other budget line item.

Table 14-1 shows the cash flow for a non-financed project[11]. Assume the project costs $100,000 and saves $28,000 per year for 15 years. This project could get approved *IF the client has $100,000 in cash to fund it.* The project has a net present value of $ 102,700 and an Internal Rate of Return of 27%.

Now, let's look at financing the project with a simple loan. Let's say the client finances the $ 100,000 for 15 years at 10% per year. That means that instead of investing $100,000 upfront (the bank provides these funds), the client pays $13,147 each year to the bank for 15 years. At the end of 15 years, the bank loan is paid off (just like a mortgage or car payment—just a different time period). *To keep this simple, ignore interest tax deductions as well as depreciation—which would likely improve the financials even further.* (See Table 14-2.)

In this case, the project generates $14,853 each year for the client. Because there is no upfront investment required, the IRR value becomes infinity.

Getting credit for your "good" energy/environmental work is often not done well by engineers. There are some good ways to express the environmental benefits (of your projects) in ways that your CEO (and the public) can appreciate. For example, instead of saying a project will save "X" kilowatt-hours, show that these avoided kWh represent avoided power plant emissions that are equivalent to planting "Y" trees. People can visualize trees, but have a hard time visualizing what a ton of CO_2 emissions looks like. To help you quantify the environmental benefits of your energy-saving projects (barrels of oil not consumed, etc.) you can download a free spreadsheet at profitablegreensolutions.com[18]. Figure 14-1 is a screen shot of the spreadsheet.

EOY	Savings	Cost	Cash Flow
0	-	(100,000)	$ (100,000)
1	28,000		$ 28,000
2	28,000		$ 28,000
3	28,000		$ 28,000
4	28,000		$ 28,000
5	28,000		$ 28,000
6	28,000		$ 28,000
7	28,000		$ 28,000
8	28,000		$ 28,000
9	28,000		$ 28,000
10	28,000		$ 28,000
11	28,000		$ 28,000
12	28,000		$ 28,000
13	28,000		$ 28,000
14	28,000		$ 28,000
15	28,000		$ 28,000
NPV i=10%			$102,700
IRR			27%

Table 14-1. Flow (paid with Cash)

However, there are even more benefits… when you consider the following impacts the project could have on:
- Shareholders in the Annual Report,
- Community Morale & "Green Image,"
- Productivity Improvements,
- Legal Risk Reduction,
- LEED Points, White Certificates, RECs[19]
- FREE Public Press[20]

Alternatively, if you want to demonstrate the environmental impact of your current operations (to get people motivated to change), there is a free carbon audit tool (FreeCarbonAudit.com) which is very helpful (see Figure 14-2).

In addition, if you have one, leverage your marketing department to get your project approved. More often than not, they will be excited about the project you are working on and it helps them develop "authentic"

EOY	Savings	Finance Cost	Cash Flow
0	-	-	$ -
1	28,000	13,147	$ 14,853
2	28,000	13,147	$ 14,853
3	28,000	13,147	$ 14,853
4	28,000	13,147	$ 14,853
5	28,000	13,147	$ 14,853
6	28,000	13,147	$ 14,853
7	28,000	13,147	$ 14,853
8	28,000	13,147	$ 14,853
9	28,000	13,147	$ 14,853
10	28,000	13,147	$ 14,853
11	28,000	13,147	$ 14,853
12	28,000	13,147	$ 14,853
13	28,000	13,147	$ 14,853
14	28,000	13,147	$ 14,853
15	28,000	13,147	$ 14,853
NPV i=10%			$112,970
IRR			n/a

Table 14-2. Financed Project Cash Flow

green marketing messages for their company, which can improve profits.

FREE Money

In addition, there are utility rebates, tax refunds, credits, and other sources of free money that will improve a project's financial return. Here are some useful websites that allow you to see utility and tax benefits in your state:

www.dsireusa.org/
www.energytaxincentives.org
http://www.efficientbuildings.org
http://www.lightingtaxdeduction.org/

But don't just rely on websites. Use professionals—they should know what techniques, technologies and rebates are best for your geographic area.

Figure 14-1.

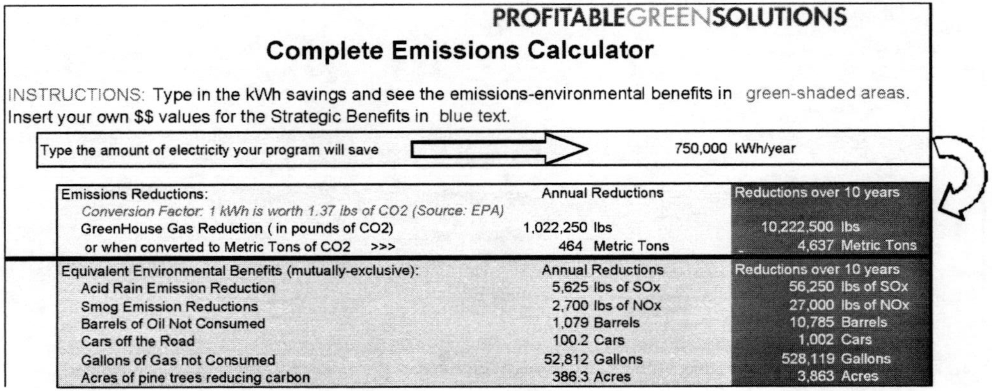

PROFITABLEGREENSOLUTIONS

Complete Emissions Calculator

INSTRUCTIONS: Type in the kWh savings and see the emissions-environmental benefits in green-shaded areas.
Insert your own $$ values for the Strategic Benefits in blue text.

Type the amount of electricity your program will save	750,000 kWh/year	

Emissions Reductions:	Annual Reductions	Reductions over 10 years
Conversion Factor: 1 kWh is worth 1.37 lbs of CO2 (Source: EPA)		
GreenHouse Gas Reduction (in pounds of CO2)	1,022,250 lbs	10,222,500 lbs
or when converted to Metric Tons of CO2 >>>	464 Metric Tons	4,637 Metric Tons
Equivalent Environmental Benefits (mutually-exclusive):	Annual Reductions	Reductions over 10 years
Acid Rain Emission Reduction	5,625 lbs of SOx	56,250 lbs of SOx
Smog Emission Reductions	2,700 lbs of NOx	27,000 lbs of NOx
Barrels of Oil Not Consumed	1,079 Barrels	10,785 Barrels
Cars off the Road	100.2 Cars	1,002 Cars
Gallons of Gas not Consumed	52,812 Gallons	528,119 Gallons
Acres of pine trees reducing carbon	386.3 Acres	3,863 Acres

Figure 14-2.

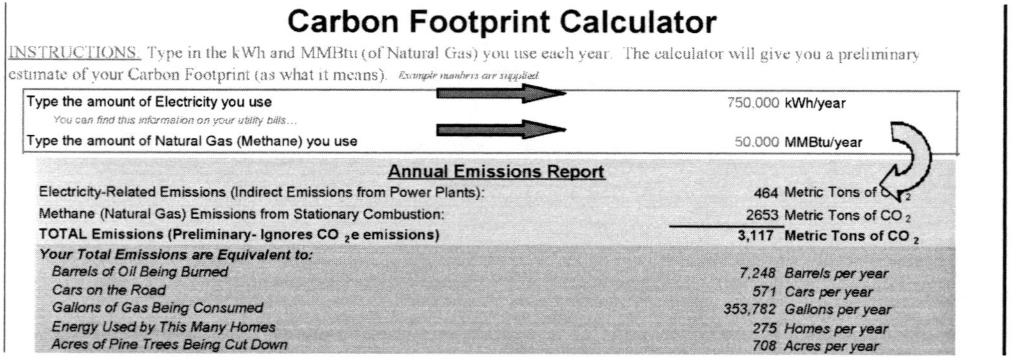

Carbon Footprint Calculator

INSTRUCTIONS: Type in the kWh and MMBtu (of Natural Gas) you use each year. The calculator will give you a preliminary estimate of your Carbon Footprint (as what it means). *Example numbers are supplied*

Type the amount of Electricity you use
You can find this information on your utility bills....   750,000 kWh/year

Type the amount of Natural Gas (Methane) you use   50,000 MMBtu/year

Annual Emissions Report

Electricity-Related Emissions (Indirect Emissions from Power Plants):	464 Metric Tons of CO_2
Methane (Natural Gas) Emissions from Stationary Combustion:	2653 Metric Tons of CO_2
TOTAL Emissions (Preliminary- Ignores CO_2e emissions)	**3,117 Metric Tons of CO_2**

Your Total Emissions are Equivalent to:

Barrels of Oil Being Burned	*7,248 Barrels per year*
Cars on the Road	*571 Cars per year*
Gallons of Gas Being Consumed	*353,782 Gallons per year*
Energy Used by This Many Homes	*275 Homes per year*
Acres of Pine Trees Being Cut Down	*708 Acres per year*

SUMMARY

This article has described the three common barriers (marketing, education and money) as well as a start on how to overcome them. To get a project approved:

1. Articulate the Problem/Pain
2. Collaborate to Add Value in the Solution
3. Quantify all the Benefits
4. Minimize Financial Risk
5. Develop/Configure an Executive Summary that "sings" to the hearts of the approver.

Hopefully, these techniques will help you get your next project approved. Why is this important... because the ice is melting! We are counting on you.

Footnotes

1. Ultimate Marketing for Engineers Course, www.ProfitableGreenSolutions.com
2. Wingender, J. and Woodroof, E., (1997) "When Firms Publicize Energy Management Projects: Their Stock Prices Go Up"—How much—21.33% on Average! *Strategic Planning for Energy and the Environment*, Summer Issue 1997.
3. The "Vault Files," www.ProfitableGreenSolutions.com
4. Download the FREE emissions calculator from www.ProfitableGreenSolutions.com
5. Other benefits can be found in this article: Woodroof, E., Tuner, W. and Heinz, S. (2008), "The Secret Benefits from Energy Conservation Contribute Value Worth An 18% Improvement To Energy Savings," *Strategic Planning for Energy and the Environment*, Vol 28(1).
6. For Example: an energy-efficient project that saves $100,000 in operating costs is equivalent to generating $1,000,000 in new sales (assuming the company has a 10% profit margin). It can be more difficult to add $1,000,000 in sales, and would require more infrastructure, etc..
7. Several examples include: Patagonia, Google, GE, Home Depot, etc.. Other examples can be downloaded from the "Resource Vault" at www.ProfitableGreenSolutions.com
8. For Example: a construction firm switched to hybrid vehicles and also offset the carbon emissions. The firm's name is prominently displayed on each vehicle. They get tons of new business because they are seen and known as the "greenest construction firm" in the city. Plus, they charge a premium for their services!
9. For Example: a law firm renovated their office in a "green" manner and attracted a new client (who chose the firm due to its "green" emphasis). The new client was worth an extra $100,000 in revenue in the first month.
10. Additional Examples: "Green" networking groups such as "greendrinks.org" and can supplement the traditional business networking clubs like Rotary Club, Kiwanis, Chamber of Commerce, etc. Also, when joining groups such as the Climate Action Registry, companies are exposed to other members, who could be superior suppliers, clients and partners.

11. Marketing to Millions Manual, Bob Circosta Communications, LLC.
12. LEED = Leadership in Energy & Environmental Design
13. Results from the Profitable Green Strategies Course, www.ProfitableGreenSolutions.com
14. www.GreenTravelPartners.com
15. MARR= Minimum Attractive Rate of Return. For more info on this topic see: Woodroof, E., Thumann, A.(2005) *Handbook for Financing Energy Projects*, Fairmont Press, Atlanta.
16. U.S. Department of Energy, (1996) "Analysis of Energy-Efficiency Investment Decisions by Small and Medium-Sized Manufacturers," U.S. DOE, Office of Policy and Office of Energy Efficiency and Renewable Energy, pp. 37-38.
17. Advanced Project Financing Course, www.ProfitableGreenSolutions.com
18. See "Resources Tab" at www.ProfitableGreenSolutions.com
19. REC = Renewable Energy Credit
20. Press release samples from the "Vault" at www.ProfitableGreenSolutions.com

Index